Rhyacophilidae of India

Rhyacophilidae of India

Sajad Hussain Parey
Tabraq Ali
Manpreet Singh Pandher

JENNY STANFORD
PUBLISHING

Published by

Jenny Stanford Publishing Pte. Ltd.
101 Thomson Road
#06-01, United Square
Singapore 307591

Email: editorial@jennystanford.com
Web: www.jennystanford.com

British Library Cataloguing-in-Publication Data
A catalogue record for this book is available from the British Library.

Rhyacophilidae of India

Cover image: Species *Rhyacophila kando rengma.*

ISBN 978-981-4968-74-4 (Hardcover)
ISBN 978-1-003-41095-9 (eBook)

Contents

Foreword

John C. Morse, Ph.D.
Professor Emeritus of Entomology
Director Emeritus of The Clemson University Arthropod Collection
Clemson University, Clemson, SC 29634-0310, USA

Caddisflies are best known for their ability to act as sensitive bioindicators, serving a critical role in the biomonitoring of water quality. They provide essential ecosystem services; for example, besides their critical role in biomonitoring, they access a wide range of nutrients in sustainable freshwater food webs, become food themselves for many aquatic and terrestrial animals, and help stabilize stream substrates.

The family Rhyacophilidae is very diverse, with about 860 species in the northern hemisphere, of which most have been documented from the Oriental region. The free-living larvae of this family are predators that help regulate populations of other insects and serve as food themselves for larger predators.

Concerning the disproportionate threats against freshwater biota worldwide, conservation strategies are urgently needed to safeguard the species of rhyacophilids and the crucial caddisfly fauna globally. Although Caddisfly researchers are scarce in the world in general and in India in particular, I am delighted to know that the concern for discovering and protecting these important animals has reached every part of the world.

This book, *Rhyacophilidae of India*, is a marvelous contribution to the world's knowledge of Trichoptera. Scientific literature is scarce and scattered in India; this tome provides all known data for this family assembled in a single book. Besides the good quality of newly drawn figures, the figures of Schmid, Kimmins, and Morton are refined, which makes the book more attractive and valuable. The field experiences shared by the authors will motivate young researchers to investigate caddisflies and other aquatic insects.

This book is unique, providing current, updated literature for all the Rhyacophilidae species recorded in the region and will be useful for ecologists investigating the particular functional traits of these different species. It will serve as a valuable resource and guide for professionals, researchers, and students. I greatly appreciate the efforts by the authors in completing this impressive project and wish them and this book every success.

John C. Morse

John C. Morse

May 27, 2022

Preface

Caddisflies are the sister taxon to Lepidoptera, the butterfly and moth order, which together form the superorder Amphiesmenoptera. Trichoptera adults, like those of Lepidoptera, are terrestrial but have setae rather than scales on their wings. Their color is somber, an adaptation that allows them to conceal themselves in foliage during the day. With nearly 17,000 species, it is the seventh most speciose order of all insects, with more species than the combined total of all the other primarily aquatic insect orders (Ephemeroptera, Odonata, Plecoptera, Megaloptera). The larvae of most Trichoptera are recognized by their transportable cases and fixed shelters. With a few exceptions, the eggs, larvae, and pupae are completely reliant on the aquatic environment and can be found in a wide range of freshwater environments, including springs, mountain streams, rivers, waterfalls, ponds, temporary waters, and wave-washed shorelines and still bottoms of deep lakes. Pupae live in shelters constructed by mature larvae, sometimes also within semipermeable cocoons under the shelters; mature pupae are covered with a semitransparent pupal integument with the pharate adult visible inside.

Caddisflies are effective bio-indicators because they tend to reflect the surrounding abiotic environment as well as changes in that habitat over time. These are often found only in higher-quality streams and so have low Tolerance Values (TVs). TVs can differ among different families, genera, and species; most Trichoptera are extremely pollutant sensitive, with low TVs. Their presence or absence indicates unpolluted versus polluted water bodies. Caddisflies are also crucial in maintaining ecological balance. Many fishes and other water, terrestrial, and aerial animals feed on the larvae, pupae, and adults. Caddisfly communities are a marker of a healthy ecosystem and indicate the level of pollution in streams.

Larvae of species in the family Rhyacophilidae are free-living, without a case or retreat. They are voracious predators of other insects, frequenting cool, fast-flowing rivers and streams. When mature, they create a domed pupal chamber of pebbles over a semipermeable cocoon.

The Rhyacophilidae family is quite speciose and is one of the most diverse of the 51 Trichoptera families, with more than 860 species worldwide and approximately 184 species and eight subspecies in two genera in India and surrounding areas. This family of caddisflies primarily inhabits north temperate and subtropical latitudes, with populations primarily in North America, Europe, and Asia, but also in India and southeastern Asia's subtropical zones. In India, the Himalayan region dominates in terms of the number of caddisfly species in general and of species in the family Rhyacophilidae in particular.

In India, Rhyacophilidae literature is scarce. There is relatively little information available on the Internet. Fernand Schmid published the majority of work in French. Kimmins, Morton, Martynov, Malicky, Saini, and Kaur are among the others.

We assembled information about the entire family in the form of a book for this project so that future generations can access the literature in one resource. Each species' taxonomic description is detailed along with good-quality figures. For each species, information on the holotype has been provided. Synonymy (if any), a thorough description, the holotype depository, additional material examined, the distribution, the geographic status, and the diagnostic characteristics are all included in the descriptive account of each species.

A detailed and updated checklist of the caddisfly family Rhyacophilidae in India, keys for the genera, groups, and subgroups have also been incorporated. The methods of collection, preservation and identification are given with illustrations.

Sajad Hussain Parey

Chapter 1

Introduction

Insects are incredibly effective animals that, despite their small size, have a significant impact on many parts of our life. Insect communities in all forms of natural and artificial habitats, both terrestrial and aquatic, typically sustain a staggering array of lifestyles, forms and functions. Because of their amazingly great species richness, insects are gaining in popularity as a scientific issue due to their current high visibility. Despite all that, some of the aquatic insect orders remained intentionally ignored at the hands of naturalists over the globe. The order Trichoptera Kirby, 1813 is one of the largest aquatic insect orders exceeding 1000 species on record from the Indian subcontinent. However, this record is yet incomplete as there are still more than 4000 species awaiting their discovery Schmid (1984). This is mainly because innumerable habitats are yet to be explored and have great constraints in collecting and procuring the scattered literature.

Caddisflies' scientific name (Trichos = setae/hair, ptera = wing) relates to the setae/hair coating on the wing surface. The insects belonging to this order are commonly called Caddisflies/Casefiles/Watermoths or even Watermotors, and their larvae as the Caddice worms because of their ability to construct and inhabit within their cases. Caddisfly adults are generally brown, grey, or tan in color, holometabolous, small to medium-sized (1.5–40 mm), slender and soft-bodied moth-like insects, which can be easily identified by their typical sitting posture i.e., while at rest, wings are

Rhyacophilidae of India
Sajad Hussain Parey, Tabraq Ali, and Manpreet Singh Pandher

ISBN 978-981-4968-74-4 (Hardcover), 978-1-003-41095-9 (eBook)
www.jennystanford.com

held on an oblique vertical plane and the antennae usually about as long as forewings.

Trichoptera is especially significant as the sister taxon of one of the megadiverse insect orders, Lepidoptera, and together they form the superorder Amphiesmenoptera, or "dressed-up wings," which refers to the extensive coverage of scales or hair on the wings. Both morphological and molecular evidence strongly supports the monophyly of these two orders (Wheeler et al., 2001). Trichoptera has hair instead of scales, which gives them a more primitive appearance. Adult Caddisflies, unlike moths and butterflies, which have a coiled, tubelike proboscis for feeding, lack well-developed mouthparts, including mandibles, but have an advanced haustellum (synapomorphic for the order) developed from a fusion of the hypopharynx and labium, which is used to ingest liquids in some species. Caddisfly adults, unlike Lepidoptera (butterflies and relatives), have a dull coloring that helps them blend in with their surroundings. The adults, larvae and pupae of the Caddisflies are frequently imitated by fly-fishers using artificial lures to emulate caddisfly emergencies during trout fishing. The Trichopteran larvae, with a few exceptions, are aquatic and primarily detritivorous unlike the Lepidopteran larvae, which are predominantly terrestrial herbivores. Caddisfly larvae, like lepidopterans, have the tendency to weave silk from specially adapted salivary glands. Trichoptera larvae are well recognized for their transportable cases and fixed shelters, which many, but not all, species build. Due to the silk secretion, Trichoptera larvae have developed a vast array of morphological modifications to cope with life in practically any type of freshwater ecology thanks to silk (Wiggins, 1996). Although the etymology of the name is unknown, it is assumed to come from "cadaz" or "cadace" (caddys), a word with varying spellings used in Shakespeare's times to refer to a ribbon fashioned from a specific type of yarn sold by traveling salesmen, who were known as "Cadice men" as a result (Hickin, 1967).

As per the data available in the "Trichoptera World Checklist," it is the seventh most speciose order of all insects, with 16,267 living species worldwide, more than the combined total of all other exclusively aquatic insect orders. The 632 genera and 63 families of caddisfly species are divided into two suborders: Annulipalpia and Integripalpia. There are also 521 fossil species, with some belonging to 133 fossil genera and 20 fossil families

(Trichoptera World Checklist, accessed on March 02, 2022). There are 265 fossilized caddisfly cases (ichnospecies) among these fossils, which have been classified into 11 ichnogenera but not to any particular family (Morse et al., 2019). The Trichoptera World Checklist divides the world into eight biogeographic regions: the Antarctic, Afrotropical (sub-Saharan Africa), Australasian (Australia, New Zealand, New Guinea, and smaller southwest Pacific islands), East Palearctic (Northern Asia), Nearctic (North America and northern Mexico), Neotropical (Southern Mexico, Central America, and South America), Oriental (Southern Asia) and West Palearctic (Europe and Mediterranean region) (Morse et al., 2019). New species are being described at a rapid rate, and it appears that the predictions of Schmid (1984) and Flint et al. (1999) are coming true, notably in the Neotropics, Madagascar, moist regions of Africa, Southeast Asia, China, and the Philippines. Only two regions, the Oriental and Neotropical, have documented more than half of the known species (de Moor and Ivanov, 2008). The Oriental region has the greatest diversity of species. Except for the Neotropics, it has more than 3,700 species, which is more than double the number of species documented in any of the other regions (de Moor and Ivanov, 2008). In India, the order Trichoptera includes 27 families, 102 genera, and about 1300 species (Morse, Personal communication 2021).

Figure 1.1 (A–D) Pictures showing examples of Caddisflies' habitat and field surveys across the Himalaya.

Family Rhyacophilidae

The family Rhyacophilidae is the most primitive and has proved to be of unusual interest biogeographically. The adults of the family Rhyacophilidae are called primitive Caddisflies. Rhyacophilidae Stephens, 1836, is a larger family of Integripalpia suborder, with Rhyacophila Pictet, 1834, as the type genus. Ross (1956) and Schmid (1970) updated the taxonomy of this family in India. Caddisflies of this family are mostly found in north-temperate countries, such as North America, Europe, and Asia, although their distribution also extends southwards into India and southeastern Asia's tropical zones.

The larvae are predatory, free-living, and prefer to live in cool, fast-flowing rivers and streams. Abdominal and thoracic gills are common in larvae. They differ from larvae of the suborder Annulipalpia by having a sclerite on abdominal tergum IX. They are distinguished from larvae of other Integripalpia suborder families by the absence of mesonotal sclerites and a prosternal horn (Holzenthal et al., 2007).

Rhyacophilidae pupae, like those of other Integripalpia basal lineages, are encased in a semipermeable cocoon beneath their domed pupal shelter made of tiny rocks. The pupae lack apical terminal lobes and slender sclerotized projections; abdominal terga III, IV, and V each possess two pairs of sclerotized hook plates; and the mandibles have preapical teeth. Adults of Rhyacophilidae have three dorsal ocelli on their heads. The sclerotization of the apical segment of each 5-segmented maxillary palp is not fragmented or annulated, and the second segment of each 5-segmented maxillary palp is short and spherical. Each foretibia also has a preapical spur in addition to the apical pair (Ali et al., 2020).

The Rhyacophilidae family is represented in India by two genera: *Rhyacophila* Pictet, 1834 and *Himalopsyche* Banks, 1940. *R. vulgaris* Pictet, 1834 is the type species of the genus *Rhyacophila* (selected by Ross, 1944). It has a predominantly Holarctic and Oriental distribution, with 793 species worldwide. Among these, 432 species have been documented from the Oriental region, while 165 species have been recorded from India and its near surroundings (Ali et al., 2020). The *Himalopsyche* genus is believed to have evolved in the Oriental region (Schmid, 1989). This genus has 50 species worldwide, with one species from the Nearctic, nine from the East Palearctic, and 43 from the Oriental region (Morse, 2020). *Himalopsyche tibetana* (Martynov, 1930) is the type species for *Himalopsyche* (originally *Rhyacophila tibetana*). Around 19 species among the 43 species of *Himalopsyche* of Oriental are found in India and its surrounding territories (Kaur and Saini, 2015).

Chapter 2

Taxonomic Attributes for Delimiting Rhyacophilidae Family

Various morphological characters such as presence or absence of ocelli, number of segments of maxillary palp, number of segments of labial palp, setal warts, tibial spurs, wing venation, and genitalia are most reliable characters for discrimination of various taxa within this insect order. While dealing with Caddisfly taxonomy in general and systematic of Rhyacophilidae particular, the following characters have been found trustworthy and stable.

1. **Ocelli (Fig. 2.1A):** The presence or absence of three ocelli, which can be distinguished by their bead like-appearance, is an important character for splitting the families into two groups. Ocelli are present in the family under consideration.
2. **Maxillary palp (Fig. 2.1B):** The number and relative length of the maxillary palp is an important taxonomic character. In Rhyacophilidae, the number of segments is five in both the sexes. In Rhyacophilidae, the first two segments are very short, second subglobular, and last segment ending in a small point.
3. **Labial palp (Fig. 2.1B):** It is always three segmented lying next to maxillary palp.
4. **Thoracic notal warts (Figs. 2.1E–F):** The thoracic warts which are of considerable value in separating different families of the Caddisflies are the pitted elevations on the dorsum of the thorax

Rhyacophilidae of India
Sajad Hussain Parey, Tabraq Ali, and Manpreet Singh Pandher

ISBN 978-981-4968-74-4 (Hardcover), 978-1-003-41095-9 (eBook)
www.jennystanford.com

and are often more setose than the surrounding area. They vary in size, number and arrangement on pro, meso and metathorax. These characters help in distinguishing the genera of family Rhyacophilidae.

5. **Tibial spurs (Fig. 2.1D):** The tibiae are equipped with spurs whose number is constant and stable character that can help in distinguishing closely related genera and species. This number is generally expressed in the form of a formula known as spur formula which indicates the number of spurs on fore, middle and hind tibiae. For example, the family Rhaycophilidae is separated from Hydrobiosidae in having 3,4,4 spurs in the former and 2,4,4 in latter.
6. **Wings:** Wings provide a wealth of taxonomic characters which include shape, venation and various other aspects. The wings may be oval or rounded at apex; forewing uniformly colored or speckled with light or dark spots. Arrangement, course and number of veins play the most important role, the application of these characters is so wide and multifacet that it can be safely applied in distinguishing various genera and species.
7. **Genitalia:** The broad structural lines of genitalia provide excellent characters for the classification of genera, while the details of the shape provide essential characters for the diagnosis of the species.

   (a) **Male genitalia:** The main parts which are generally considered are:

      (i) Tergite VIII: The shape and indentation of tergite VIII is an important character for distinguishing the species.

      (ii) Segment IX: Dorsally, in some cases segment IX is modified as apical lobe. Whereas ventrally it bears one or two segmented inferior appendages.

      (iii) Segment X: Segment X may bear preanal appendage which is completely or partially free or fused. Its shape may be oval, rounded or truncate at apex. Anal sclerite is a modification of segment X and its presence or absence also play important role in identifying the species.

      (iv) Phallic apparatus: Various parts of the phallic apparatus like dorsal appendage of phallic apparatus, phallicata and paramere are of considerable importance while distinguishing closely related species. Paramere may be spiniform or glabrous, and its shape is one of the most constant and conservative character that can be exploited at the species level.

(b) Female genitalia: In the female, there are 11 abdominal segments, all of which are well developed. The characters which lead to the determination of a species are as follows:

(i) Segment VIII: Segment VIII shows remarkable diversity and play an important role in identifying species.

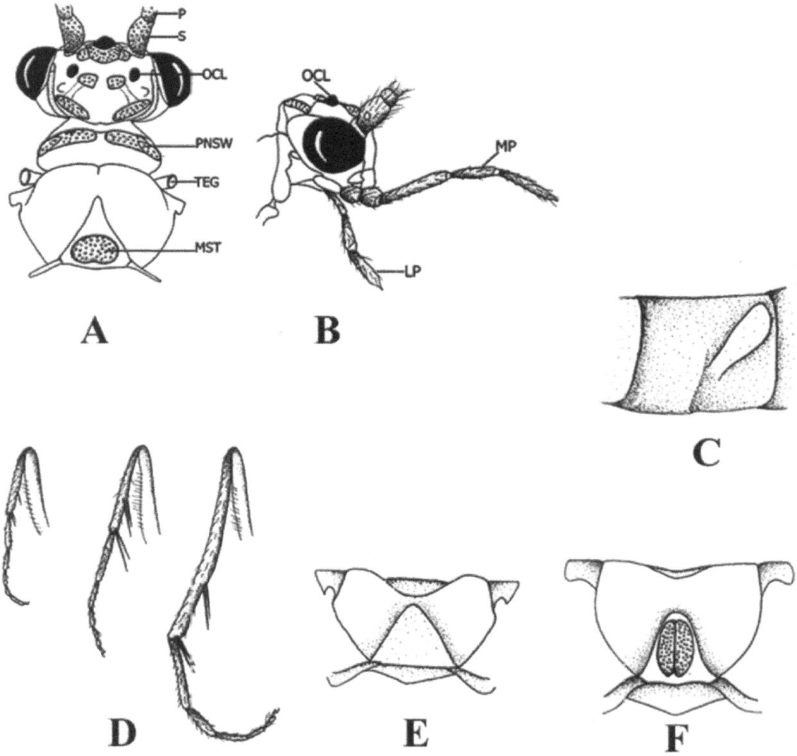

Figure 2.1 (A) Head and thorax of Rhyacophilidae (*Rhyacophila obscura*, ♂), dorsal view. (B) Lateral view of head of Rhyacophilidae (*Rhyacophila obscura*, ♂). (C) Sternite V of *Rhyacophila obscura* (♂). (D) Foreleg, midleg and hindleg of Rhyacophilidae (*Rhyacophila obscura*, ♂). (E) Metathorax of *Rhyacophila obscura* (♂). (F) Metathorax of *Himalopsyche angnorbui* (♀).

(ii) Segment IX: Shape of lateral margins of segment IX is an important and constant species-specific character.

(iii) Segment X: Segment X may be fused with segment XI or completely free from it. It may be conical, triangular, rounded or indented at apex.

(iv) Vaginal apparatus: Shape of posterior process, intermediate process and processus spermathecae of vaginal apparatus help in discrimination of species.

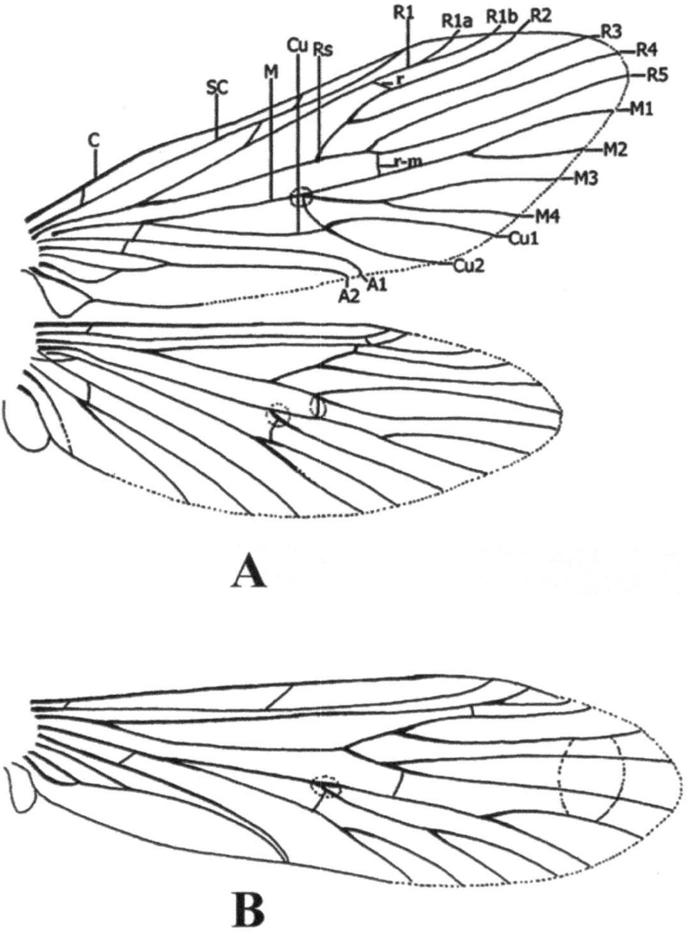

Figure 2.2 (A) Forewing and hindwing of *Rhyacophila lepcha* (♂). (B) Forewing of *Himalopsyche horai* (♂).

Chapter 3

Methods of Collection, Preservation, and Identification

3.1 Light Trap

The primary methods of the collection of adult caddisflies are done using fluorescent blue light and mercury bulb with 12-volt rechargeable batteries placed across a white sheet (Figs. 3.1A–B). These collections are made near the streams, tributaries, and the main rivers with varying water quality, temperature, and altitude ponds, freshwater lakes from 1–4 hours from dusk. The majority of caddisflies emerge and begin flying at dusk, or shortly before dark, and continue to fly to the sheet for another 2–3 hours after sunset. After that, activity is irregular, but adults maintain a low level of activity throughout the night, with a secondary peak just before dawn. The caddisflies are generally recognized in the field due to their sitting posture (wings held in a triangular roof over the abdomen) and continuously vibrating long antennae. The collection of specimens may be dry or wet; in dry collection, the adults are collected directly from the light trap or stones near the freshwater into a jar containing a few drops of ethyl acetate in a cotton (for killing purpose). In the wet collection, the adults are collected directly from the trap into a jar containing 70–96% ethanol.

Rhyacophilidae of India
Sajad Hussain Parey, Taoraq Ali, and Manpreet Singh Pandher

ISBN 978-981-4968-74-4 (Hardcover), 978-1-003-41095-9 (eBook)
www.jennystanford.com

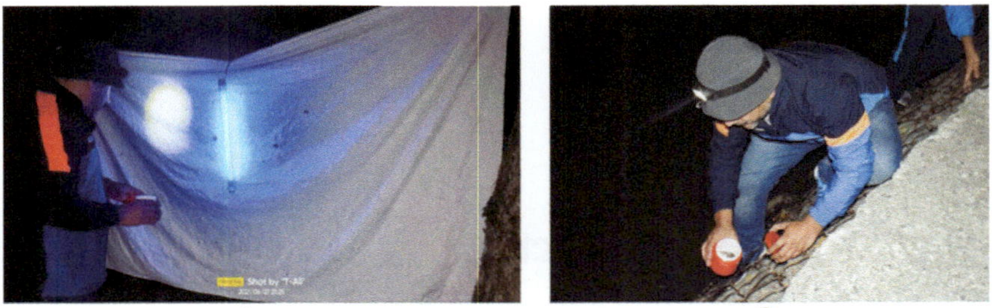

A B

Figure 3.1 (A) Light trap with fluorescent blue light powered by a 12 V battery. (B) Collection of Caddisflies from stones near the stream.

3.2 Pan Trap/Bucket Light Trap

Another option is to apply a passive methodology (pan trap/bucket light trap method), such as fixing a lantern or fluorescent bulb on or near a tray (Blahnik & Holzenthal, 2004). The most basic pan light trap is made up of an illumination device (often a fluorescent bulb) and a collecting recipient with alcohol or another preservation solution at the bottom. The insects that are attracted to the light are caught in the trap and swiftly die from the alcohol. The light pan trap is powered by 12 V batteries that power a fluorescent daylight bulb (15 W/12 V) and a UV bulb (15 W/12 V) (Figs. 3.2A–B).

A B

Figure 3.2 (A) The components of light pan trap (Adolfo Ricardo Calor & Rodolfo Mariano, 2009). (B) Bucket light trap.

3.3 Malaise Trap

Malaise traps are an important adjunct collecting technique, particularly for a large number of day-active species and those that are not easily attracted to lights. Malaise traps function best if they can be set up for several days and checked for specimens on a regular basis. These traps are set up over tiny, first-order woodland streams and serve both as flight intercept and emergence traps. Adults are trapped in the trap's fluid-filled reservoir (which contains ethyl alcohol or an ethylene glycol solution), then extracted and transported to the requisite percent of ethanol for permanent storage (Fig. 3.3).

Figure 3.3 Malaise trap.

3.4 Aerial/Sweeping Net

Some species of the genus *Rhyacophila* may also be collected from aquatic vegetation during daytime with the help of sweeping nets. During the day, many species can be seen actively displaying on or above native vegetation or rocks. Some show off by crawling along the coast or in the stream. Adult activity is most common in the morning when the tropical sun is just warming the habitat and the afternoon rains have not yet arrived. This is an excellent time to walk along a section of stream with a sweep net in hand in order to catch day-active insects (Fig. 3.4).

Figure 3.4 Daylight collection using aerial net.

A B

Figure 3.5 (A–B) Collection of Caddisflies larvae near a waterbody.

For preservation, the specimens are transferred into the vials by putting in 70% or 96% ethanol for morphological and molecular studies. For the molecular analysis, the specimens must be preserved in 96% Ethanol so that the DNA may not be degraded. These specimens are stored in –20°C.

The larvae of caddisflies are usually collected from streams using kick nets, hand nets, and hand-picking methods (Fig. 3.5A–B).

3.4.1 Identification

For the identification of caddisflies, the examination of various morphological characters at different levels is carried out. General morphological features are used only up to the family and generic level. Species are sorted out on the basis of the external morphology of genitalia only as the general morphology of other body parts plays a minor or negligible role in species determination. The identification is done under the stereozoom binocular and trinocular microscopes (Fig. 3.6). For the line diagrams, the inking of the final drawings is done with the Rottering black ink.

Figure 3.6 Identification of Caddisflies using stereozoom binocular and trinocular microscopes.

The terminology proposed by Nielsen (1957, 1980) for the genitalia of both the sexes may be adopted. The dichotomous keys of the families as proposed by Saini *et al.* (2001) and for species identification, the works of Schmid (1959, 1963, 1966, 1970), Morton (1900), Martynov (1935), Kimmins (1952, 1953) and Saini and Kaur (2011, 2012, 2013, 2015) are adopted in India. To examine genitalia, male and female genitalia are prepared for observation by removing the abdominal tip with fine-tipped forceps. After removing, the genitalia are placed in KOH (10–12 %) overnight. After taking out the genitalia from KOH, it is put in alcohol so that organic matter inside may become clear. The lactic acid procedure as outlined by Blahnik *et al.* (2007) is another method for clearing genitalia.

For female genitalia, after treating the material with lactic acid procedure or KOH, a syringe full of alcohol is used to flush the abdomen. This involves inserting the syringe in the open anterior end of the abdomen, holding the abdomen and syringe with forceps, and then flushing the alcohol from the syringe into the abdomen. This generally flushes the soft tissues and extends the abdomen for viewing.

Chapter 4

Rhyacophilidae Larvae

Trichoptera larvae are most famous for the moveable cases and permanent shelters that many, but not all, species make. These larvae can be identified from all other insects with segmented thoracic legs, by having a pair of anal prolegs, each with a single curved terminal a claw and very short, sometimes almost invisible single-segmented antenna. The pupae are exarate and covered by a semitransparent integument and if fully developed reveal the pharate adult inside. The pupae usually possess a pair of strong functional mandibles (nonfunctional in adults) and the abdomen with a number of segments adorned with characteristic sclerotized, dorsal hook-bearing plates. With a few exceptions, Trichoptera larvae and pupae are completely reliant on water habitats water habitats and can be found in a wide range of freshwater ecosystems, including springs, mountain streams, big rivers, the splash locations of waterfalls, swampy wetlands, along coastlines and in the depths of lakes, as well as temporary waters. They are apneustic and dependent on dissolved oxygen for respiration. Habitat diversity among different species of Caddisflies ranges from cold, swift, mountain torrents with a maximum temperature of 45° to 55 °F to sluggish rivers and lakes with temperatures in excess of 70 °F. Trichoptera occurs in all faunal regions except Antarctica (Wiggins, 1982).

The larvae of Rhyacophilidae reside in cold streams, do not create retreats, and do not have a case i.e., they are free living.

Rhyacophilidae of India
Sajad Hussain Parey, Tabraq Ali, and Manpreet Singh Pandher

ISBN 978-981-4968-74-4 (Hardcover), 978-1-003-41095-9 (eBook)
www.jennystanford.com

The majority of species can be found in mountainous areas with plenty of cold, fast-moving streams. They are distinguished by unsclerotized meso and metanotum, sclerotized ninth abdominal tergum, and lengthy anal prolegs (Bouchard, 2004) (Fig. 4.1). Most species' life cycles are generally univoltine, but some may be semivoltine. Most species' larvae are active predators, although a few species' larvae feed on living and dead vascular plant tissue as well as algae (Weaver and Morse, 1986).

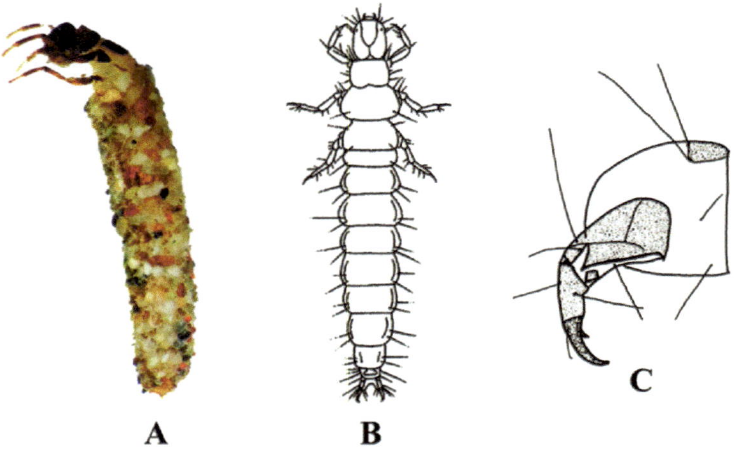

Figure 4.1 (A) Larval case. (B) Lateral view of the *Rhyacophila* larva. (C) The terminal abdominal segment of the *Rhyacophila* larva (lateral view), shows anal proleg.

Chapter 5

Updated Checklist of Family Rhyacophilidae in India

I. Genus *Himalopsyche* Banks, 1940

1. *Himalopsyche amitabha* Schmid, 1966

   Distribution: India (Sikkim)

   Geographical position: Native to Indian Himalaya
2. *Himalopsyche angnorbui angnorbui* Schmid, 1963

   Distribution: Nepal, India (Uttarakhand, Sikkim, Himachal Pradesh)

   Geographical position: Vastly dispersed Himalayan species

   Subspecies: *Himalopsyche angnorbui sherpa* Schmid, 1963

   Distribution: India (Sikkim)

   Geographical position: Native to Indian Himalaya
3. *Himalopsyche bhagirathi* Schmid, 1963

   Distribution: Nepal, India (Uttarakhand)

   Geographical position: Vastly dispersed Himalayan species
4. *Himalopsyche biansata* Kimmins, 1952

   Distribution: Tibet, Bhutan, India (Sikkim)

   Geographical position: Vastly dispersed Himalayan species
5. *Himalopsyche digitata* (Martynov, 1935), *Rhyacophila*

   Distribution: Bhutan, Nepal, India (Arunachal Pradesh, Sikkim,

Rhyacophilidae of India
Sajad Hussain Parey, Tabraq Ali, and Manpreet Singh Pandher

ISBN 978-981-4968-74-4 (Hardcover), 978-1-003-41095-9 (eBook)
www.jennystanford.com

West Bengal, Uttarakhand, Sikkim) Geographical position: Vastly dispersed Himalayan species

6. *Himalopsyche dolmasampa* Schmid, 1963

   Distribution: Nepal, India (Sikkim, Uttarakhand)

   Geographical position: Vastly dispersed Himalayan species

7. *Himalopsyche gyamo* Schmid, 1963

   Distribution: Nepal, India (Sikkim)

   Geographical position: Vastly dispersed Himalayan species

8. *Himalopsyche hierophylax* Schmid, 1966

   Distribution: India (Uttarakhand)

   Geographical position: Native to Indian Himalaya

9. *Himalopsyche horai* (Martynov, 1936), *Rhyacophila*

   (Syn. *Rhyacophila pallida* Martynov, 1935)

   (Syn. *Himalopsyche phedongensis* Kimmins, 1952)

   Distribution: Bhutan, Nepal, India (Arunachal Pradesh, Himachal Pradesh, Sikkim, Uttarakhand)

   Geographical position: Vastly dispersed Himalayan species

10. *Himalopsyche lanceolata* (Morton, 1900), *Rhyacophila*

    Distribution: Myanmar, India (Meghalaya, Manipur)

    Geographical position: Vastly dispersed Himalayan species

11. *Himalopsyche lepcha* Schmid, 1963

    Distribution: Bhutan, India (Sikkim, West Bengal)

    Geographical position: Vastly dispersed Himalayan species

12. *Himalopsyche lungma* Schmid, 1963

    Distribution: India (Uttarakhand)

    Geographical position: Native to Indian Himalaya

13. *Himalopsyche maitreya* Schmid, 1963

    Distribution: Nepal, India (Uttarakhand, Arunachal Pradesh)

    Geographical position: Vastly dispersed Himalayan species

14. *Himalopsyche malenanda* Schmid, 1963

    Distribution: Bhutan, Nepal, India (Arunachal Pradesh, Sikkim, Uttarakhand)

    Geographical position: Vastly dispersed Himalayan species

15. *Himalopsyche schmidi* Kaur and Saini, 2015
    Distribution: India (Himachal Pradesh)
    Geographical position: Native to Indian Himalaya
16. *Himalopsyche tibetana* (Martynov, 1930), *Rhyacophila*
    Distribution: Bhutan, Nepal, Tibet, India (Sikkim, Uttarakhand)
    Geographical position: Vastly dispersed Himalayan species
17. *Himalopsyche todma* Schmid, 1963
    Distribution: Nepal, India/Pakistan (Uttarakhand, Jammu & Kashmir, Himachal Pradesh)
    Geographical position: Vastly dispersed Himalayan species
18. *Himalopsyche yatrawalla* Schmid, 1966
    Distribution: India (Uttarakhand)
    Geographical position: Native to Indian Himalaya
19. *Himalopsyche yongma* Schmid, 1963
    Distribution: India (Sikkim)
    Geographical position: Native to Indian Himalaya

II. Genus *Rhyacophila* F. J. Pictet, 1834

1. *Rhyacophila acutis* Kaur & Saini, 2012
   Distribution: India (Meghalaya)
   Geographical position: Native to Indian Himalaya
2. *Rhyacophila alticola* Kimmins, 1953
   Distribution: Nepal, Tibet, India (Sikkim)
   Geographical position: Vastly dispersed Himalayan species
3. *Rhyacophila anatina* Morton, 1900
   Distribution: China, India (Meghalaya, Sikkim)
   Geographical position: Vastly dispersed Himalayan species
4. *Rhyacophila ancestralis* Martynov, 1935
   Distribution: India (Meghalaya)
   Geographical position: Native to Indian Himalaya
5. *Rhyacophila angden* Schmid, 1970
   Distribution: India (Sikkim)
   Geographical position: Native to Indian Himalaya

6. *Rhyacophila antiope* Malicky, 1997
   Distribution: India (North Andaman Island)
   Geographical position: An Island species endemic to India
7. *Rhyacophila armitageorum* Saini & Kaur, 2012
   Distribution: India (Arunachal Pradesh)
   Geographical position: Native to Indian Himalaya
8. *Rhyacophila assimilis* Kimmins, 1953
   Distribution: Myanmar, India (Manipur)
   Geographical position Vastly dispersed Himalayan species
9. *Rhyacophila aureomaculata* Schmid, 1970
   Distribution: India (Arunachal Pradesh)
   Geographical position: Native to Indian Himalaya
10. *Rhyacophila aureostigma* Schmid, 1970
    Distribution: Bhutan, Nepal, India (West Bengal)
    Geographical position: Vastly dispersed Himalayan species
11. *Rhyacophila bhotia* Schmid, 1970
    Distribution: India (Uttarakhand)
    Geographical position: Native to Indian Himalaya
12. *Rhyacophila bhuchanadhara* Schmid, 1970
    Distribution: Bhutan, India (Arunachal Pradesh)
    Geographical position: Vastly dispersed Himalayan species
13. *Rhyacophila bicolor* Kimmins, 1953
    Distribution: Myanmar, Vietnam, India (Manipur)
    Geographical position: Vastly dispersed Himalayan species
14. *Rhyacophila bidens* Kimmins, 1953
    (Syn. *Rhyacophila szaboi* Kiss, 2013)
    Distribution: Myanmar, Bhutan, Nepal, Vietnam, Pakistan/ India (Arunachal Pradesh, Manipur, Sikkim, Uttarakhand, Jammu & Kashmir)
    Geographical position: Vastly dispersed Himalayan species
15. *Rhyacophila chakungpa* Schmid, 1970
    Distribution: Nepal, India (Sikkim)
    Geographical position: Vastly dispersed Himalayan species

16. *Rhyacophila chamolungpa* Schmid, 1970
    Distribution: India (Arunachal Pradesh)
    Geographical position: Native to Indian Himalaya
17. *Rhyacophila chandzo* Schmid, 1970
    Distribution: Nepal, India (Arunachal Pradesh)
    Geographical position: Vastly dispersed Himalayan species
18. *Rhyacophila changpa* Schmid, 1970
    Distribution: Nepal, India (Sikkim)
    Geographical position: Vastly dispersed Himalayan species
19. *Rhyacophila chayulpa chayulpa* Schmid, 1970
    Distribution: Bhutan, Nepal, India (Sikkim, Uttarakhand)
    Geographical position: Vastly dispersed Himalayan species
    Subspecies: *Rhyacophila chayulpa ringmo* Schmid, 1970
    Distribution: India (Arunachal Pradesh)
    Geographical position: Native to Indian Himalaya
    Subspecies: *Rhyacophila chayulpa tsetangpa* Schmid, 1970
    Distribution: India (Uttarakhand)
    Geographical position: Native to Indian Himalaya
20. *Rhyacophila chematangpa* Schmid, 1970
    Distribution: India (Sikkim)
    Geographical position: Native to Indian Himalaya
21. *Rhyacophila chembo chembo* Schmid, 1970
    Distribution: Nepal, Bhutan, India (Arunachal Pradesh)
    Geographical position: Vastly dispersed Himalayan species
    Subspecies: Rhyacophila *chembo lartsepa* Schmid, 1970
    Distribution: India (Uttarakhand)
    Geographical position: Native to Indian Himalaya
22. *Rhyacophila chenmo* Schmid, 1970
    Distribution: India (Manipur)
    Geographical position: Native to Indian Himalaya
23. *Rhyacophila chimdro* Schmid, 1970
    Distribution: India (Arunachal Pradesh)
    Geographical position: Native to Indian Himalaya

24. *Rhyacophila chomoyuma* Schmid, 1970
    Distribution: India (Manipur)
    Geographical position: Native to Indian Himalaya
25. *Rhyacophila choprai* Martynov, 1935
    Distribution: India (Uttarakhand)
    Geographical position: Native to Indian Himalaya
26. *Rhyacophila chugalungpa* Schmid, 1970
    Distribution: India (Sikkim)
    Geographical position: Native to Indian Himalaya
27. *Rhyacophila chulukpa* Schmid, 1970
    Distribution: India (West Bengal)
    Geographical position: Native to Indian Himalaya
28. *Rhyacophila chumikpa* Schmid, 1970
    Distribution: India (Arunachal Pradesh)
    Geographical position: Native to Indian Himalaya
29. *Rhyacophila chungse* Schmid, 1970
    Distribution: India (Arunachal Pradesh, Manipur, Nagaland)
    Geographical position: Native to Indian Himalaya
30. *Rhyacophila churongpa* Schmid, 1970
    Distribution: India (Uttarakhand)
    Geographical position: Native to Indian Himalaya
31. *Rhyacophila curvata* (Morton, 1900)
    (Syn. *Rhyacophila mortoni* Kimmins, 1953)
    (Syn. *Rhyacophila sumatrana* Ulmer, 1930)
    (Syn. *Rhyacophila suthepensis* Malicky, 1987)
    Distribution: Thailand, China, Indonesia, Malaysia, Myanmar, India (Meghalaya, Manipur, Sikkim)
    Geographical position: Vastly dispersed Himalayan species
32. *Rhyacophila dafla* Schmid, 1970
    Distribution: India (Arunachal Pradesh)
    Geographical position: Native to Indian Himalaya

33. *Rhyacophila dakshi* Schmid, 1970

    Distribution: Nepal, India (West Bengal)

    Geographical position: Vastly dispersed Himalayan species

34. *Rhyacophila dgaldanpa* Schmid, 1970

    Distribution: India (Arunachal Pradesh)

    Geographical position: Native to Indian Himalaya

35. *Rhyacophila dikkaravasini* Schmid, 1970

    Distribution: India (Chennai)

    Geographical position: Native to South India

36. *Rhyacophila dilata* Martynov, 1935

    Distribution: Nepal, India (Himachal Pradesh, Uttarakhand, Sikkim)

    Geographical position: Vastly dispersed Himalayan species

37. *Rhyacophila dirangpa* Schmid, 1970

    Distribution: India (Arunachal Pradesh)

    Geographical position: Native to Indian Himalaya

38. *Rhyacophila dolingpa* Schmid, 1970

    Distribution: India (Meghalaya)

    Geographical position: Native to Indian Himalaya

39. *Rhyacophila dongkyapa* Schmid, 1970

    Distribution; Nepal, Bhutan, India (Arunachal Pradesh)

    Geographical position: Vastly dispersed Himalayan species

40. *Rhyacophila dongre* Schmid, 1970

    Distribution: India (Sikkim)

    Geographical position: Native to Indian Himalaya

41. *Rhyacophila dorje* Schmid, 1970

    Distribution: India (Chennai)

    Geographical position: Native to South India

42. *Rhyacophila drokpa drokpa* Schmid, 1970

    Distribution: Thailand, Bhutan, India (Arunachal Pradesh, Uttarakhand)

    Geographical position: Vastly dispersed Himalayan species

    Subspecies: *Rhyacophila drokpa nyenpa* Schmid, 1970

Distribution: India (Sikkim)

Geographical position: Native to Indian Himalaya

43. *Rhyacophila drosampa* Schmid, 1970

    Distribution: Thailand, India (Meghalaya)

    Geographical position: Vastly dispersed Himalayan species

44. *Rhyacophila drotangpa* Schmid, 1970

    Distribution: India (Sikkim)

    Geographical position: Native to Indian Himalaya

45. *Rhyacophila extensa* Martynov, 1927

    (Syn. *Rhyacophila carletoni* Banks, 1931)

    Distribution: Turkestan (Kirghizstan), India (Himachal Pradesh)

    Geographical position: Vastly dispersed Himalayan species

46. *Rhyacophila fletcheri* (Kimmins, 1952)

    (Syn. *Trichophila fletcheri* Kimmins, 1952)

    Distribution: India (Arunachal Pradesh, Sikkim, Uttarakhand)

    Geographical position: Native to Indian Himalaya

47. *Rhyacophila gelukpa* Schmid, 1970

    Distribution: Vietnam, India (Arunachal Pradesh, Sikkim)

    Geographical position: Vastly dispersed Himalayan species

48. *Rhyacophila gyaldzen* Schmid, 1970

    Distribution: India (Meghalaya)

    Geographical position: Native to Indian Himalaya

49. *Rhyacophila gyamo* Schmid, 1970

    Distribution: Thailand, India (Manipur)

    Geographical position: Vastly dispersed Himalayan species

50. *Rhyacophila gyaspa* Schmid, 1970

    Distribution: India (West Bengal)

    Geographical position: Native to Indian Himalaya

51. *Rhyacophila gyelbu* Schmid, 1970

    Distribution: India (Manipur)

    Geographical position: Native to Indian Himalaya

52. *Rhyacophila himachalensis* Kaur & Saini, 2012
    Distribution: India (Himachal Pradesh)
    Geographical position: Native to Indian Himalaya
53. *Rhyacophila himalayensis* Saini, Kaur and Bajwa, 2013
    Distribution: India (Uttarakhand)
    Geographical position: Native to Indian Himalaya
54. *Rhyacophila hingstoni* Martynov, 1930
    Distribution: Nepal, Tibet, India (Sikkim)
    Geographical position: Vastly dispersed Himalayan species
55. *Rhyacophila hobsoni* Martynov, 1930
    Distribution: Nepal, Tibet, India (Sikkim, Uttarakhand)
    Geographical position: Vastly dispersed Himalayan species
56. *Rhyacophila inconspicua* Morton, 1900
    Distribution: India (Meghalaya)
    Geographical position: Native to Indian Himalaya
57. *Rhyacophila indica* Saini, Kaur and Bajwa, 2012
    Distribution: India (Sikkim)
    Geographical position: Native to Indian Himalaya
58. *Rhyacophila jayadurga* Schmid, 1970
    Distribution: India (Chennai)
    Geographical position: Native to Indian Himalaya
59. *Rhyacophila kadampa* Schmid, 1970
    Distribution: India (Manipur)
    Geographical position: Native to Indian Himalaya
60. *Rhyacophila kadaphes* Schmid, 1959
    Distribution: Pakistan/India (Jammu & Kashmir, Uttarakhand, Himachal Pradesh)
    Geographical position: Native to Indian Himalaya/Pakistan
61. *Rhyacophila kagyupa* Schmid, 1970
    Distribution: India (Arunachal Pradesh)
    Geographical position: Native to Indian Himalaya

62. *Rhyacophila kando kando* Schmid, 1970

    Distribution: Nepal, Bhutan, India (Uttarakhand, Arunachal Pradesh)

    Geographical position: Vastly dispersed Himalayan species

    Subspecies: *Rhyacophila kando rengma* Schmid, 1970

    Distribution: India (Manipur)

    Geographical position: Native to Indian Himalaya

63. *Rhyacophila kangjongpa* Schmid, 1970

    Distribution: India (Arunachal Pradesh)

    Geographical position: Native to Indian Himalaya

64. *Rhyacophila kanichka* Schmid, 1959

    Distribution: Pakistan/India (Uttarakhand, Jammu & Kashmir)

    Geographical position: Native to Indian Himalaya/Pakistan

65. *Rhyacophila karpa* Schmid, 1970

    Distribution: India (Meghalaya)

    Geographical position: Native to Indian Himalaya

66. *Rhyacophila kashmirensis* Kaur and Saini, 2013

    Distribution: Pakistan/India (Jammu & Kashmir)

    Geographical position: Native to Indian Himalaya/Pakistan

67. *Rhyacophila kashongpa* Schmid, 1970

    Distribution: India (Manipur)

    Geographical position: Native to Indian Himalaya

68. *Rhyacophila kawachenpa* Schmid, 1970

    Distribution: India (Arunachal Pradesh)

    Geographical position: Native to Indian Himalaya

69. *Rhyacophila kedara* Schmid, 1970

    Distribution: Pakistan/India (Jammu & Kashmir, Uttarakhand)

    Geographical position: Native to Indian Himalaya/Pakistan

70. *Rhyacophila khamakhya* Schmid, 1970

    Distribution: India (Chennai)

    Geographical position: Native to South India

71. *Rhyacophila khasiorum* Schmid, 1970

    Distribution: Nepal, India (Himachal Pradesh, Meghalaya, Uttarakhand)

    Geographical position: Vastly dispersed Himalayan species

72. *Rhyacophila khimbarpa* Schmid, 1970

    Distribution: India (Arunachal Pradesh)

    Geographical position: Native to Indian Himalaya

73. *Rhyacophila khiympa* Schmid, 1970

    Distribution: Bhutan, China, India (Sikkim)

    Geographical position: Vastly dispersed Himalayan species

74. *Rhyacophila kubra* Schmid, 1970

    Distribution: Bhutan, Nepal: India (Sikkim)

    Geographical position: Vastly dispersed Himalayan species

75. *Rhyacophila kunma* Schmid, 1970

    Distribution: India (Arunachal Pradesh)

    Geographical position: Native to Indian Himalaya

76. *Rhyacophila kusang* Schmid, 1970

    Distribution: India (Meghalaya)

    Geographical position: Native to Indian Himalaya

77. *Rhyacophila kyadongpa* Schmid, 1970

    Distribution: India (Sikkim)

    Geographical position: Native to Indian Himalaya

78. *Rhyacophila kyimdongpa* Schmid, 1970

    Distribution: Vietnam, Thailand, India (Meghalaya)

    Geographical position: Vastly dispersed Himalayan species

79. *Rhyacophila kyungpa* Schmid, 1970

    Distribution: India (Arunachal Pradesh)

    Geographical position: Native to Indian Himalaya

80. *Rhyacophila langdarma* Schmid, 1970

    Distribution: India (Arunachal Pradesh)

    Geographical position: Native to Indian Himalaya

81. *Rhyacophila laptsapa* Schmid, 1970

    Distribution: Nepal, India (Sikkim)

    Geographical position: Native to Indian Himalaya

82. *Rhyacophila lepcha* Schmid, 1970

    Distribution: Bhutan, India (Uttarakhand, Sikkim)

    Geographical position: Vastly dispersed Himalayan species

83. *Rhyacophila lhabu* Schmid, 1970

    Distribution: Vietnam, India (Manipur)

    Geographical position: Vastly dispersed Himalayan species

84. *Rhyacophila lhadzongpa* Schmid, 1970

    Distribution: Pakistan/India (Jammu & Kashmir, Arunachal Pradesh, Himachal Pradesh)

    Geographical position: Native to Indian Himalaya/Pakistan

85. *Rhyacophila lhakpa* Schmid, 1970

    Distribution: India (Himachal Pradesh, Uttarakhand)

    Geographical position: Native to Indian Himalaya

86. *Rhyacophila lhopa* Schmid, 1970

    Distribution: India (Chennai)

    Geographical position: Native to South India

87. *Rhyacophila lobsang* Schmid, 1970

    Distribution: India (Chennai)

    Geographical position: Native to South India

88. *Rhyacophila lonpo* Schmid, 1970

    Distribution: India (Arunachal Pradesh)

    Geographical position: Native to Indian Himalaya

89. *Rhyacophila maitripa* Schmid, 1970

    Distribution: India (Arunachal Pradesh)

    Geographical position: Native to Indian Himalaya

90. *Rhyacophila manipuri* Schmid, 1970

    Distribution: India (Manipur)

    Geographical position: Native to Indian Himalaya

91. *Rhyacophila manlungpa* Schmid, 1970
    Distribution: India (Arunachal Pradesh)
    Geographical position: Native to Indian Himalaya
92. *Rhyacophila marpa* Schmid, 1970
    Distribution: India (Arunachal Pradesh)
    Geographical position: Native to Indian Himalaya
93. *Rhyacophila milarepa* Schmid, 1970
    Distribution: India (Arunachal Pradesh)
    Geographical position: Native to Indian Himalaya
94. *Rhyacophila mishmica* Kimmins, 1953
    Distribution: India (Arunachal Pradesh)
    Geographical position: Native to Indian Himalaya
95. *Rhyacophila monyulpa* Schmid, 1970
    Distribution: India (Arunachal Pradesh)
    Geographical position: Native to Indian Himalaya
96. *Rhyacophila muktepa* Schmid, 1970
    Distribution: Thailand, India (Meghalaya)
    Geographical position: Vastly dispersed Himalayan species
97. *Rhyacophila nabochepa* Schmid, 1970
    Distribution: India (West Bengal)
    Geographical position: Native to Indian Himalaya
98. *Rhyacophila naga* Schmid, 1970
    Distribution: India (Manipur)
    Geographical position: Native to Indian Himalaya
99. *Rhyacophila nagongpa* Schmid, 1970
    Distribution: India (Meghalaya)
    Geographical position: Native to Indian Himalaya
100. *Rhyacophila nakpo* Schmid, 1970
    Distribution: India (Manipur)
    Geographical position: Native to Indian Himalaya
101. *Rhyacophila namgyal* Schmid, 1970
    Distribution: India (Meghalaya)
    Geographical position: Native to Indian Himalaya

102. *Rhyacophila narayani* Schmid, 1970

 Distribution: India (Kerala)

 Geographical position: Native to South India

103. *Rhyacophila naviculata* Morton, 1900

 Distribution: India (Meghalaya, Tamil Nadu)

 Geographical position: A widely distributed species endemic to India

104. *Rhyacophila netongpa* Schmid, 1970

 Distribution: India (Arunachal Pradesh)

 Geographical position: Native to Indian Himalaya

105. *Rhyacophila ngawang* Schmid, 1970

 Distribution: India (Meghalaya)

 Geographical position: Native to Indian Himalaya

106. *Rhyacophila ngorpa* Schmid, 1970

 Distribution: Nepal, Bhutan, Tibet, India (Sikkim)

 Geographical position: Vastly dispersed Himalayan species

107. *Rhyacophila ngulpa* Schmid, 1970

 Distribution: Nepal, India (Himachal Pradesh, Uttarakhand)

 Geographical position: Vastly dispersed Himalayan species

108. *Rhyacophila nigrorosea* Schmid, 1959

 Distribution: Pakistan, India (Uttarakhand, Himachal Pradesh)

 Geographical position: Vastly dispersed Himalayan species

109. *Rhyacophila norbu* Schmid, 1970

 Distribution: India (Meghalaya)

 Geographical position: Native to Indian Himalaya

110. *Rhyacophila nyamangpa* Schmid, 1970

 Distribution: India (Sikkim)

 Geographical position: Native to Indian Himalaya

111. *Rhyacophila nyelungpa* Schmid, 1970

 Distribution: India (Meghalaya)

 Geographical position: Native to Indian Himalaya

112. *Rhyacophila nyerongpa* Schmid, 1970

Distribution: India (Arunachal Pradesh)

Geographical position: Native to Indian Himalaya

113. *Rhyacophila nyerpa* Schmid, 1970

Distribution: India (Arunachal Pradesh)

Geographical position: Native to Indian Himalaya

114. *Rhyacophila nyukmadongpa* Schmid, 1970

Distribution: India (Arunachal Pradesh)

Geographical position: Native to Indian Himalaya

115. *Rhyacophila obscura* Martynov, 1927

Distribution: Bhutan, Turkestan (Uzbekistan), Afghanistan, Nepal, Pakistan/India (Sikkim, Himachal Pradesh, Arunachal Pradesh, Jammu & Kashmir, Uttarakhand)

Geographical position: Vastly dispersed Himalayan species

116. *Rhyacophila parva parva* Kimmins, 1953

Distribution: India (Meghalaya)

Geographical position: Native to Indian Himalaya

Subspecies: *Rhyacophila parva mukpo* Schmid, 1970

Distribution: India (Manipur)

Geographical position: Native to Indian Himalaya

117. *Rhyacophila paurava* Schmid, 1959

Distribution: Pakistan, India (Uttarakhand, Meghalaya)

Geographical position: Native to Indian Himalaya/Pakistan

118. *Rhyacophila pemba* Schmid, 1970

Distribution: India (Meghalaya)

Geographical position: Native to Indian Himalaya

119. *Rhyacophila poba* Schmid, 1970

Distribution: Tibet, India (Sikkim)

Geographical position: Vastly dispersed Himalayan species

120. *Rhyacophila polha* Schmid, 1970

Distribution: India (Arunachal Pradesh)

Geographical position: Native to Indian Himalaya

121. *Rhyacophila procliva* Kimmins, 1953

Distribution: Myanmar, India (Arunachal Pradesh, Manipur)

Geographical position: Vastly dispersed Himalayan species

122. *Rhyacophila putata putata* Kimmins, 1953

Distribution: India (Meghalaya)

Geographical position: Native to Indian Himalaya

Subspecies: *Rhyacophila putata temba* Schmid, 1970

Distribution: India (Manipur)

Geographical position: Native to Indian Himalaya

123. *Rhyacophila rhombica* Martynov, 1935

Distribution: Vietnam, Bhutan, Nepal, India (West Bengal, Himachal Pradesh)

Geographical position: Vastly dispersed Himalayan species

124. *Rhyacophila rongpa* Schmid, 1970

Distribution: Nepal, Bhutan, India (Arunachal Pradesh, Manipur, Uttarakhand)

Geographical position: Vastly dispersed Himalayan species

125. *Rhyacophila sajadi* Saini & Kaur, 2012

Distribution: India (Arunachal Pradesh, Uttarakhand, Himachal Pradesh, Sikkim, Assam)

Geographical position: Native to Indian Himalaya

126. *Rhyacophila sakyapa* Schmid, 1970

Distribution: Nepal, India (Uttarakhand)

Geographical position: Vastly dispersed Himalayan species

127. *Rhyacophila sanglungpa* Schmid, 1970

Distribution: India (Sikkim)

Geographical position: Native to Indian Himalaya

128. *Rhyacophila scissa* Morton, 1900

Distribution: Myanmar, Bhutan, China, Vietnam, Pakistan/ India (Sikkim, Arunachal Pradesh, Uttarakhand, Jammu & Kashmir, Meghalaya, Himachal Pradesh, Nagaland)

Geographical position: Vastly dispersed Himalayan species

129. *Rhyacophila scissoides* Kimmins, 1953

Distribution: Vietnam, Nepal, China, Thailand, India (Meghalaya, Arunachal Pradesh, Uttarakhand, Sikkim)

Geographical position: Vastly dispersed Himalayan species

130. *Rhyacophila scotina* Kimmins, 1953

Distribution: India (Meghalaya)

Geographical position: Native to Indian Himalaya

131. *Rhyacophila senggepa* Schmid, 1970

Distribution: India (Meghalaya)

Geographical position: Native to Indian Himalaya

132. *Rhyacophila shakangpa* Schmid, 1970

Distribution: Nepal, India (Uttarakhand)

Geographical position: Vastly dispersed Himalayan species

133. *Rhyacophila sherchokpa* Schmid, 1970

Distribution: India (Arunachal Pradesh)

Geographical position: Native to Indian Himalaya

134. *Rhyacophila sherpa* Schmid, 1970

Distribution: India (Sikkim)

Geographical position: Native to Indian Himalaya

135. *Rhyacophila shingripa* Schmid, 1970

Distribution: India (Sikkim)

Geographical position: Native to Indian Himalaya

136. *Rhyacophila sikungpa* Schmid, 1970

Distribution: India (Uttarakhand)

Geographical position: Native to Indian Himalaya

137. *Rhyacophila similis* Martynov, 1935

Distribution: Taiwan, India (Uttarakhand)

Geographical position: Vastly dispersed Himalayan species

138. *Rhyacophila spinalis* Martynov, 1930

Distribution: Nepal, Tibet, Bhutan, India (Sikkim, Uttarakhand, Arunachal Pradesh, Meghalaya)

Geographical position: Vastly dispersed Himalayan species

139. *Rhyacophila stenostyla* Martynov, 1930
Distribution: Tibet, Nepal, India (Uttarakhand, Sikkim)
Geographical position: Vastly dispersed Himalayan species
140. *Rhyacophila sumdopa* Schmid, 1970
Distribution: India (Manipur)
Geographical position: Native to Indian Himalaya
141. *Rhyacophila tarkiya* Schmid, 1970
Distribution: India (Sikkim)
Geographical position: Native to Indian Himalaya
142. *Rhyacophila tashepa* Schmid, 1970
Distribution: India (Sikkim)
Geographical position: Native to Indian Himalaya
143. *Rhyacophila tashidingpa* Schmid, 1970
Distribution: Nepal, India (Arunachal Pradesh)
Geographical position: Vastly dispersed Himalayan species
144. *Rhyacophila tecta* Morton, 1900
Distribution: Myanmar, Bhutan, Taiwan, India (Meghalaya, Arunachal Pradesh)
Geographical position: Vastly dispersed Himalayan species
145. *Rhyacophila tengyelingpa* Schmid, 1970
Distribution: India (Sikkim)
Geographical position: Native to Indian Himalaya
146. *Rhyacophila tolungpa* Schmid, 1970
Distribution: Bhutan, India (Arunachal Pradesh)
Geographical position: Vastly dispersed Himalayan species
147. *Rhyacophila trashipa* Schmid, 1970
Distribution: Bhutan, India (Manipur)
Geographical position: Vastly dispersed Himalayan species
148. *Rhyacophila trulungpa* Schmid, 1970
Distribution: Pakistan, India (Uttarakhand)
Geographical position: Vastly dispersed Himalayan species

149. *Rhyacophila tsering* Schmid, 1970
Distribution: India (Kerala, Meghalaya)
Geographical position: A widely distributed species endemic to India

150. *Rhyacophila tshiringpa* Schmid, 1970
Distribution: Nepal, India (Sikkim)
Geographical position: Vastly dispersed Himalayan species

151. *Rhyacophila tshogpa* Schmid, 1970
Distribution: India (Sikkim)
Geographical position: Native to Indian Himalaya

152. *Rhyacophila tsiudmarpo* Schmid, 1970
Distribution: India (Arunachal Pradesh)
Geographical position: Native to Indian Himalaya

153. *Rhyacophila tsona* Schmid, 1970
Distribution: Bhutan, India (Manipur)
Geographical position: Vastly dispersed Himalayan species

154. *Rhyacophila tsongkhapa* Schmid, 1970
Distribution: India (Sikkim)
Geographical position: Native to Indian Himalaya

155. *Rhyacophila tungkorpa* Schmid, 1970
Distribution: India (Sikkim)
Geographical position: Native to Indian Himalaya

156. *Rhyacophila tungpa* Schmid, 1970
Distribution: India (Uttarakhand)
Geographical position: Native to Indian Himalaya

157. *Rhyacophila ugyenpa* Schmid, 1970
Distribution: India (Meghalaya)
Geographical position: Native to Indian Himalaya

158. *Rhyacophila wangpo* Schmid, 1970
Distribution: India (West Bengal)
Geographical position: Native to Indian Himalaya

159. *Rhyacophila yarlungpa* Schmid, 1970

Distribution: Nepal, India (Arunachal Pradesh)

Geographical position: Vastly dispersed Himalayan species

160. *Rhyacophila yigrongpa* Schmid, 1970

Distribution: India (Sikkim)

Geographical position: Native to Indian Himalaya

161. *Rhyacophila yipung* Schmid, 1970

Distribution: India (Manipur)

Geographical position: Native to Indian Himalaya

162. *Rhyacophila yishepa* Schmid, 1970

Distribution: India (Sikkim)

Geographical position: Native to Indian Himalaya

163. *Rhyacophila yonggyapa* Schmid, 1970

Distribution: India (Manipur)

Geographical position: Native to Indian Himalaya

164. *Rhyacophila yullha* Schmid, 1970

Distribution: Bhutan, India (Sikkim)

Geographical position: Vastly dispersed Himalayan species

165. *Rhyacophila zhungpa* Schmid, 1970

Distribution: Nepal, Bhutan, India (Arunachal Pradesh, Himachal Pradesh, Sikkim)

Geographical position: Vastly dispersed Himalayan species

Chapter 6

Genus *Himalopsyche* Banks, 1940

Himalopsyche Banks, 1940: 173–220.

Type species – *Rhyacophila tibetana* Martynov, 1930. Original designation.
= *Himalophanes* Banks, 1940: 201.

Type species – *Himalopsyche* (*Himalophanes*) *anomala* Banks. Original designation.

Schmid: 1966 (Genus revised).

Diagnostic features: Spurs 3,4,4. The first and second segments of maxillary palp are short, the second subglobular, the third very long, and the fifth ending in a pointed spur. Legs long flattened, with spines, spurs long and distinct. Metascutellum with spiniferous warts and strongly spotted. Wing venation has similarity with *Rhyacophila* except, veins R and M curve backward so that in forewings R4 and R5 are on each side of the apical wing point (Fig. 2.2B). The discoidal cell opens in both wings. Ventral abdominal process lacking.

1. **Indian Species of the *H. kuldschensis* Subgroup**
    1. *H. bhagirathi* Schmid, 1963
    2. *H. dolmasampa* Schmid, 1963
    3. *H. gyamo* Schmid, 1963
    4. *H. lungma* Schmid, 1963

Rhyacophilidae of India
Sajad Hussain Parey, Tabraq Ali, and Manpreet Singh Pandher

ISBN 978-981-4968-74-4 (Hardcover), 978-1-003-41095-9 (eBook)
www.jennystanford.com

5. *H. malenanda* Schmid, 1963
6. *H. schmidi* Kaur and Saini, 2015
7. *H. todma* Schmid, 1963
8. *H. yatrawalla* Schmid, 1966
9. *H. yongma* Schmid, 1963

1. Genus *Himalopsyche* Banks, 1940

Himalopsyche

Group I
1. *Himalopsyche kuldschensis*

Subgroups
1. *Himalopsyche kuldschensis* (09 sps.)
2. *Himalopsyche angnorbui* (02 sps.)

Group II
2. *Himalopsyche tibetana*

Subgroups
1. *Himalopsyche navasi* (03 sps.)
2. *Himalopsyche tibetana* (02 sps.)
3. *Himalopsyche anomala* (03 sps.)

2. Indian Species of the *H. angnorbui* Subgroup

1. *H. amitabha* Schmid, 1966
2. *H. angnorbui* Schmid, 1963

3. Indian Species of the *H. navasi* Subgroup

1. *H. horai* (Martynov, 1936)
2. *H. lanceolata* (Morton, 1900)
3. *H. lepcha* Schmid, 1963

4. Indian species of the *H. tibetana* Subgroup

1. *H. biansata* Kimmins, 1952
2. *H. tibetana* (Martynov, 1930)

5. Indian species of *H. anomala* Subgroup

1. *H. digitata* (Martynov, 1935)
2. *H. hierophylax* Schmid, 1966
3. *H. maitreya* Schmid, 1963

Key to the Indian Genera of Rhyacophilidae Stephens, 1836

1. Metascutellum with a broad, setiferous wart, sometimes bifurcated into two (Fig. 1.2F) .... *Himalopsyche* Banks, 1940
   - Metascutellum glabrous (Fig. 1.2E) .... *Rhyacophila* Pictet, 1834

Key to species groups of genus *Himalopsyche* Banks, 1940

1. Males.. 2
   – Females.. 3
2. Inferior appendage uniarticulated........***H. kuldschensis* Group**
   – Inferior appendage biarticulated.................***H. tibetana* Group**
3. VIII segment partly or completely divided into tergum and sternum..***H. kuldschensis* Group**
   – VIII segment simple and undivided...........***H. tibetana* Group**

Description of Indian species of *Himalopsyche* Banks, 1940

1. *Himalopsyche bhagirathi* Schmid

Himalopsyche bhagirathi Schmid, 1963: 209–210

Adult: Upper body reddish brown, with darker areas. Antennae brown; length of antenna 17.46 mm, ringed red. Face and palps reddish brown. Legs long and thin; brown anterior and medial femora. Abdomen dark brown to reddish brown, paler on the ventral side. Wings are large, narrow, and slender in shape. The anterior ones have a very pale greyish background color and are dotted with brown macules. Hind wings are greyish, clear and without macules. Forewing 26.19 mm. In forewing, fork 1st arises prior to 2nd.

Genitalia (♀): (Figs. 6.1A–B). Dorsum of segment VIII larger than sternum. Segment IX was indented and segment X with triangular sclerotization. Apodeme extending anterad up to almost the entire length of the VIII segment. Segment IX with oval patches. The anterior margin of segment VIII is rounded and the posterior one is indented (Fig. 6.1A). Vaginal apparatus entire and breadth of the vaginal apparatus same as its length (Fig. 6.1B).

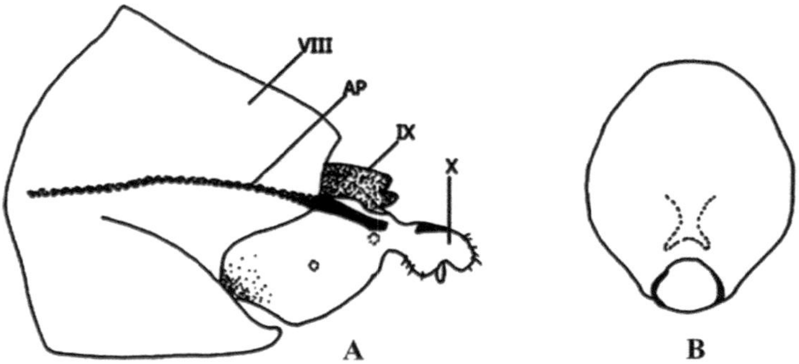

Figure 6.1 *Himalopsyche Bhagirathi* (♀): (A) Lateral view of the female genitalia; (B) Ventral view of vaginal apparatus.

Genitalia (♂): (Figs. 6.2A–B). Segment IX is strong and very elongated, slightly indented at the level of the lower appendages, with a prominent upper apical lateral angle, masking part of the Xth segment. dorsally, the IXth segment, fused in this place to a part of the X segment, forms a large dorsal appendage, in the form of a plate slightly asymmetrical, gradually widened to three-quarters of its length, then narrowed again to its end which is truncated and without a notch. The Xth segment is not very prominent, with dorsal lobes entirely distinct from each other (Fig. 6.2B). Anal scleritis is very long, slender and rather widely curved downwards. Phallotheca is not very apparent. Phallus large, in the form of a strong plate, well chitinous, very thin at its extremity which is curved downwards in a hook. Parameres are thin, very little sclerotized and regularly thinned down to the apex.

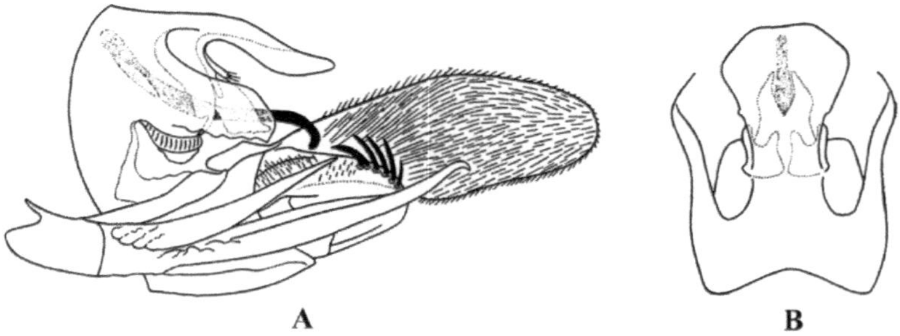

Figure 6.2 *Himalopsyche Bhagirathi* (♂): (A) Lateral view of male genitalia; (B) Dorsal view of IX segment (from Schmid, 1963).

Holotype: INDIA: Uttarakhand; Pauri Garhwal, Moussa Pani, 11000 ft., 21-VI-1958.

Holotype depository: ROM, Canada.

Additional Material examined: INDIA: Uttarakhand: Gobinddham, 3100 m, 20.vi.2011, 2♀♀. (Coll. Parey & Pandher).

Distribution: Nepal, India (Uttarakhand).

2. ***Himalopsyche dolmasampa* Schmid**

Himalopsyche dolmasampa Schmid, 1963: 212–213

Adult: *Himalopsyche dolmasampa* is a dark, heavily spotted species, inhabiting the high altitudes of the Himalaya. Upper body black-brown, with silvery areas especially in the center of the mesonotum and a fine blackish hairiness. Antennae brown, annular red. Gray-brown face. Strong, thickened and two-colored legs, like those of *malenanda*. Abdomen brown, paler below. Wings are moderately broad, but of very variable size. The forewings have a light ground color and bear numerous dark brown macules of very irregular size. In macropterous specimens, the two colors are in approximately equal proportions, while brachypterous insects are very dark and show some uniformly brown patches. Hind wings are greyish brown, a little speckled at the tip.

Genitalia (♂): (Figs. 6.3A–D). The IXth segment is very elongated all around, but strongly and obliquely notched dorsally. The Xth segment appears subovate, seen from above, with its very prominent lateral angles and a strong and long median keel. The dorsal lobes are in a relatively apical position, shaped like fingers, three times as long as they are wide and separated from each other by a fairly large space; midlobes located immediately below the dorsal lobes, fully fused to each other and forming a broad, subrectangular whole. Tergal band entirely membranous and without lateral reinforcements. Strong, thick anal scleritis with a very long root. Phallobase is short and extended obliquely upwards and backwards by strong "claspers hangers" with a circular head. The phallus is a broad, strongly chitinous plate, clearly concave upwards and terminating by a thin hook curved downwards. Paramers spiniform, chitinous, a little shorter or as long as the phallus. Lower appendages of medium size; their external lobe has the shape of a rather obtuse oval and with a sinuous dorsal edge; its internal face is covered with small, poorly developed hairs; the internal lobe of the lower appendages is a large, long triangular plate, curved upwards and rather slender end, armed with a few strong apical spines.

Genitalia (♀): Not available.

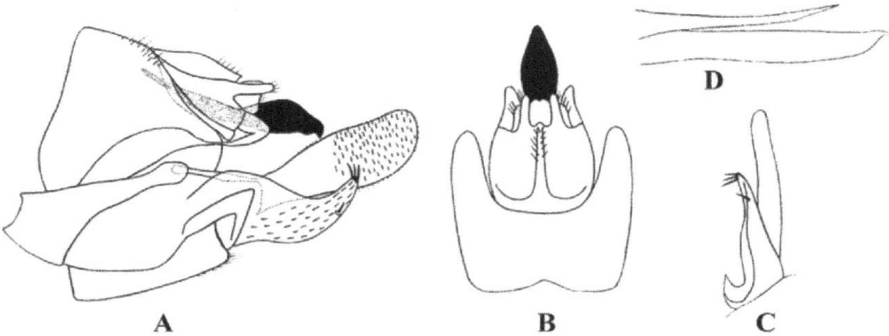

Figure 6.3 *Himalopsyche dolmasampa* (♂): (A) Lateral view of male genitalia; (B) Dorsal view of IXth and Xth segments; (C) Lower appendage, dorsal; (D) Phallic apparatus (from Schmid, 1963).

Holotype: INDIA: Sikkim; Tangshing, 12800 ft., 15-IV-1959.

Holotype depository: ROM, Canada.

Distribution: Nepal, India (Sikkim, Uttarakhand).

3. *Himalopsyche gyamo* Schmid

Himalopsyche gyamo Schmid, 1963: 218–220

Adult: Top of head and thorax reddish brown, with silvery reflections and sparse blackish hairiness. Uniformly red antennae. Face and palps reddish-gray. Uniform reddish-brown weep and paws; the legs are quite thick, and the claws strongly developed. Abdomen greyish reddish brown, a little paler below; abdominal tergites 1–7 bear very long and dense golden hair. The wings are remarkably large and narrow. The anterior ones are almost entirely hairless, except in the subthyridial region. On the hind wings, only the pterostigma is slightly tinted.

Genitalia (♂): (Figs. 6.4A–B). The IXth segment is irregularly developed; it is very elongated ventrally, shows a clear apical lateral notch, its apical dorsal angle is straight, and its apical dorsal edge does not form an overhang. The Xth segment appears subogival, seen from above, with its lateral edges reinforced but simple; at its base, it shows a hairy median prominence; its dorsal lobes are obtuse and almost completely fused to each other; the median lobes are very long and welded in a single trapezoidal piece, but with a triangular top. Anal scleritis is slightly smaller. The tergal band is simple, but with two lateral reinforcements in narrow bands bearing thin setuli. Small phallobase, with short, discoidal "claspers hangers" at the tip. Endotheca is clearly visible. The phallus is distinctly smaller and membranous, but with two longitudinal sclerotized reinforcements. Lower appendages as remarkable in size; the outer lobes (inferior appendages) are also large ovals, with their ends broadly rounded, but with their upper basal angles very open; the upper basal spinule area is smaller and the spines longer. The internal lobe is less clearly bibranched; the upper branch is obtuse and poorly separated from the lower lobe which is slenderer and bears smaller spines.

Genitalia (♀): Not available.

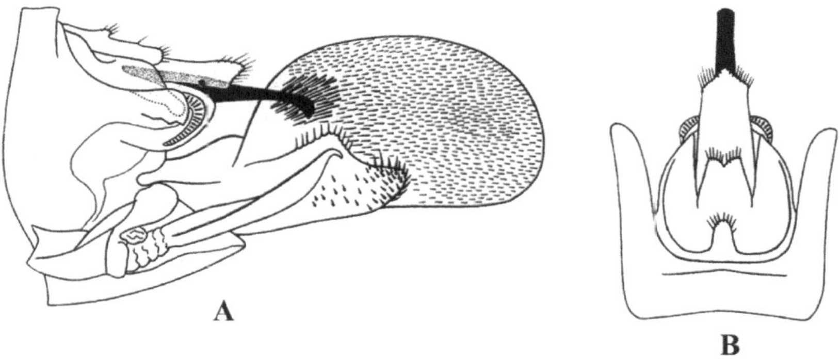

Figure 6.4 *Himalopsyche gyamo* (♂): (A) Lateral view of male genitalia; (B) Dorsal view of IXth and Xth segments (from Schmid, 1963).

Holotype: INDIA: Sikkim; Tangshing, 12800 ft., 15-IV-1959.

Holotype depository: ROM, Canada.

Distribution: Nepal, India (Sikkim).

4. *Himalopsyche lungma* Schmid

Himalopsyche lungma Schmid, 1963: 217–218

Adult: The adults of this species have a very close resemblance to *Himalopsyche gyamo*.

Genitalia (♂): (Figs. 6.5A–B). The IXth segment is regularly elongated all around, except dorsally where it is strongly narrowed and forms a short eave slightly dominating the base of the Xth segment; its apical dorsal angle is raised and prominent. The Xth segment is quite large when seen from above, showing complex lateral reinforcements very developed. Dorsal lobes are slender, finger-shaped, entirely distinct from each other and 4 to 5 times longer than wide. Mechanical lobes fused to each other over 2/3rd their length and twice as long as the dorsal lobes. Thin needle-like anal scleritis curved in a downward hook. Tergal band entirely membranous and without reinforcements. Phallobase large, rather elongated and carrying "claspers hangers" stretched and terminated in large discs. Phallotheca is clearly visible and endotheca poorly developed, Phallus appears as a strong organ, quite clearly sclerotized, flattened dorso-ventrally, clearly concave upwards and a little curved downwards at its end which is very thin. Parameres absent. Lower appendages are very remarkable for their great development. The outer lobe appears as a large oval, with the apex broadly rounded and the angle basal superior right and prominent; inside this angle, there is an area covered with strong spines directed towards the front, while the whole internal face of the piece is furnished with small clear hairs, very regularly arranged, and directed obliquely upwards. The inner lobe of the lower appendages is much smaller and bi-branched; the upper branch is triangular, quite pointed and carries small bristles; the lower branch is obtusely rounded and bears strong conical spines.

Genitalia (♀): Not available.

Holotype: INDIA: Uttarakhand; Pauri Garhwal, Jagrao, 17500 ft., 26-VI-1958.

Holotype depository: ROM, Canada.

Distribution: India (Uttarakhand).

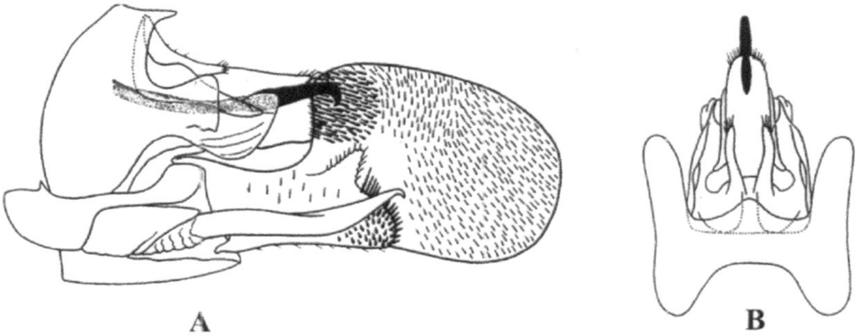

Figure 6.5 *Himalopsyche lungma* (♂): (A) Lateral view of male genitalia; (B) Dorsal view of IXth and Xth segments (from Schmid, 1963).

5. *Himalopsyche malenanda* Schmid

Himalopsyche malenanda Schmid, 1963: 210–212

Adult: Head, antenna, thorax, maxillary palp, labial palp, wings and abdomen brown. Forewing with dark brown irrotations. Foreleg and mid-leg with dark and light brown bands and hind leg pale brown. Length of antenna 13.58 mm and forewing 17.0 mm. In forewing, fork 1st arises earlier than 2nd.

Genitalia (♂): (Figs. 6.6A–C). In the dorsal view, segment IX is apically rounded and in the lateral view segment IX is curved slightly downwards. Anal sclerite hook-shaped and strongly sclerotized. Paramere is sclerotized and the inferior lobe of the inferior appendage with stout spines is smaller than the superior lobes whereas superior lobes are rounded apically. Phallicata is smaller than paramere and extends slightly beyond the IX segment posteriorly. Segment IX developed into an elongated roof that covers underlying appendages.

Genitalia (♀): Not available.

Holotype: INDIA: Sikkim; Tangshing, 12800 ft., 15-IV-1959.

Holotype depository: ROM, Canada.

Additional Material examined: INDIA: Uttarakhand: Badrinath, 3400 m, 22.vi.2011. (Coll. Parey & Pandher).

Distribution: Nepal, Bhutan, India (Arunachal Pradesh, Uttarakhand, Sikkim).

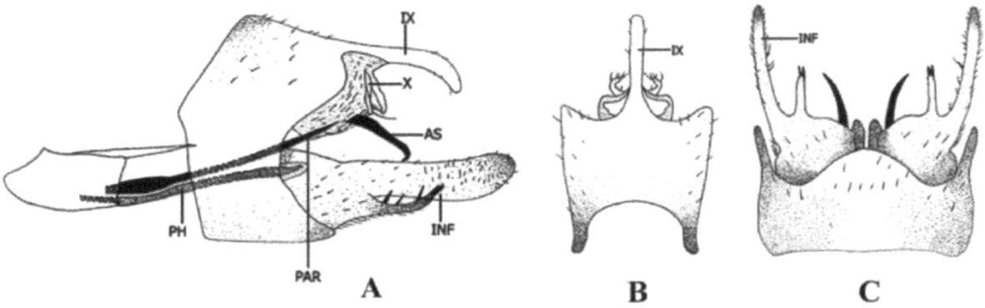

Figure 6.6 *Himalopsyche malenanda*: (A) Lateral view of male genitalia; (B) Dorsal view; (C) Ventral view.

6. ***Himalopsyche schmidi*** **Kaur and Saini**

Himalopsyche schmidi Kaur and Saini, 2015

Adult: Wings yellow and body more dark-brown than head and legs. The average length of forewing 21.0 mm and that of antenna 18.6 mm. In forewing forks 1st and 2nd arise at the same level.

Genitalia (♀): (Figs. 6.7A–B). Segment VIII is completely divided into tergite and sternite. The apex of the genitalia is rounded in outline and with indentation. Apodeme extending anterior to VII segment. The vaginal apparatus is simple in outline. The apex of the vaginal apparatus is rounded and basally truncate. Only the female specimen of *Himalopsyche schmidi* has been reported so far.

Genitalia (♀): Not available.

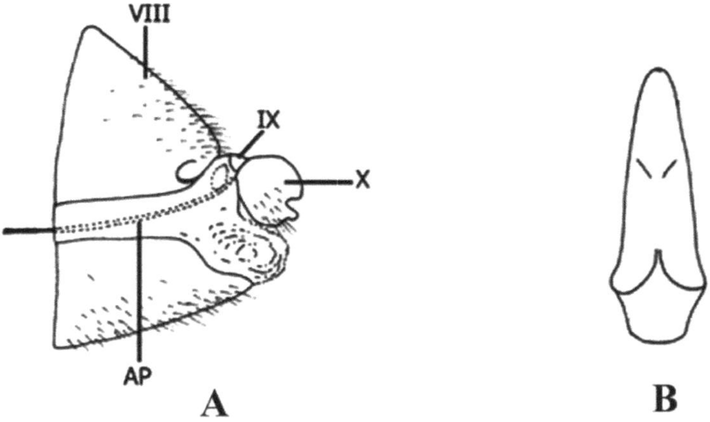

Figure 6.7 *Himalopsyche schmidi*: (A) Lateral view of female genitalia; (B) Ventral view of vaginal apparatus.

Holotype: Holotype ♀, INDIA: Himachal Pradesh: Gehra, 1200 m, 14-IX-2008. (Coll. Parey & Pandhar).

Holotype depository: MDZPUP, India.

Distribution: India (Himachal Pradesh).

7. *Himalopsyche todma* Schmid

Himalopsyche todma Schmid, 1963: 213–215.

Adult: Dark brown head brown; scape, pedicel dark brown to dark brown but flagellum palebrown; maxillary and labial palpi pale brown; thorax and abdomen brown;forewing pale brown with dark brown irrotations; hindwing hyaline; legs pale-brown but tibia of fore and midleg dark brown. The length of the antenna is 16.49 mm and that of the forewing is 23.28 mm. In forewing, fork 1st arises earlier than 2nd in both sexes. In males, the legs are orangish brown with distinct brown markings on the femur and tibia of the fore and midleg but only on the femur of the hindleg; both wings are pale brown and the forewing with brown irrotations. Length of antenna 15.52 mm and forewing 20.37 mm.

Genitalia (♀): (Figs. 6.8A–B). Tergum of segment VIII with deep indentation and sternum smooth. Segment IX is triangular in outline. Segment X rounded with terminal papilla. Apodeme extending anterad up to 2/3rd length of segment VIII. The apex of the vaginal apparatus is indented and basally divided into two parts.

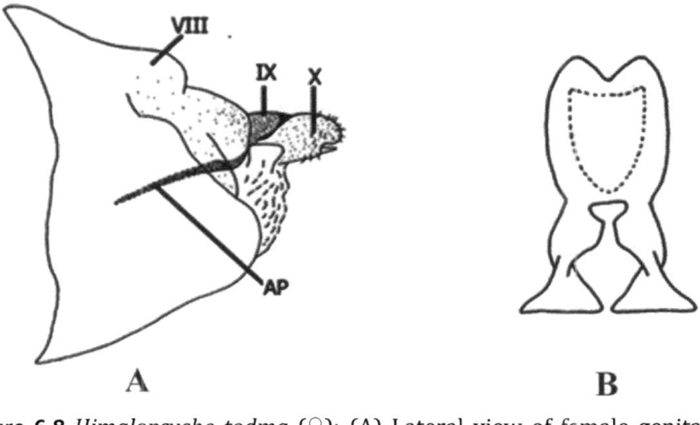

Figure 6.8 *Himalopsyche todma* (♀): (A) Lateral view of female genitalia; (B) Ventral view of vaginal apparatus.

Genitalia (♂): (Figs. 6.9A–C). In dorsal view anal sclerite is apically bifid. In the lateral view tergum of segment IX is smaller than the sternum. Sternum of segment IX three times longer than its tergum. Segment X roof-like, indented apically and slightly cover anal sclerite. Anal sclerite curved downwards. Phallicata slightly extends beyond the inferior lobe of the terminal segment of inferior appendage and curved apically. Paramere shorter than phallicata and pointed apically. Inferior lobe of inferior appendage rounded, and superior lobe raised upwards.

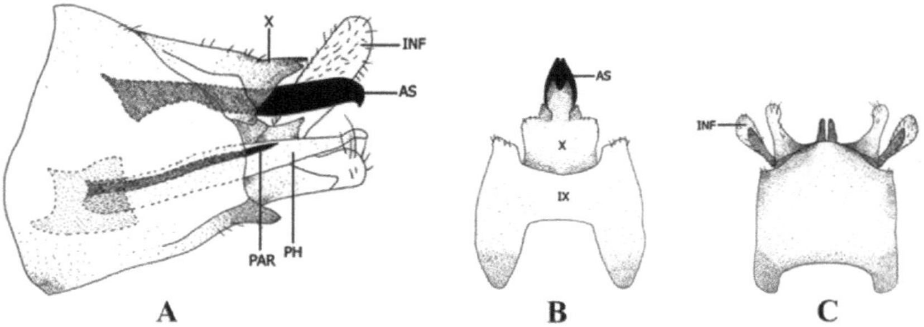

Figure 6.9 *Himalopsyche todma* (♂): (A) Lateral view of male genitalia; (B) Dorsal view; (C) Ventral view.

Holotype: INDIA: Uttarakhand; Pauri Garhwal, Phurkia, 10510 ft., 13-10-1958.

Holotype depository: ROM, Canada.

Additional material examined: INDIA: Jammu & Kashmir: Sonamarg, 3000 m, 26.vii.2011; Himachal Pradesh: Losar, 3200 m, 22.viii.2008; Chhatradu, 3600 m, 21.viii.2008; Marhi, 2700 m, 08.viii.2008; Kaza, 3000 m, 24.viii.2008; Uttarakhand: Badrinath, 3400 m, 22.vi.2012. (Coll. Parey & Pandhar).

Distribution: Nepal, Pakistan, India (Uttarakhand, Himachal Pradesh, Jammu & Kashmir).

8. *Himalopsyche yatrawalla* Schmid

Himalopsyche Yatrawalla Schmid, 1966: 143–144.

Adult: Brown and red body. Light red antennae. Forewings are not very dark, but strongly and finely speckled with light and dark brown, with two white spots, especially large, in the post-costal area.

Genitalia (♂): (Figs. 6.10A–D). The male genitalia of *Himalopsyche yatrawalla* is easily recognizable by the dorsal vestigial lobe of the IXth segment. The anal sclerites are relatively small and the internal lobe of the lower appendages long and thin.

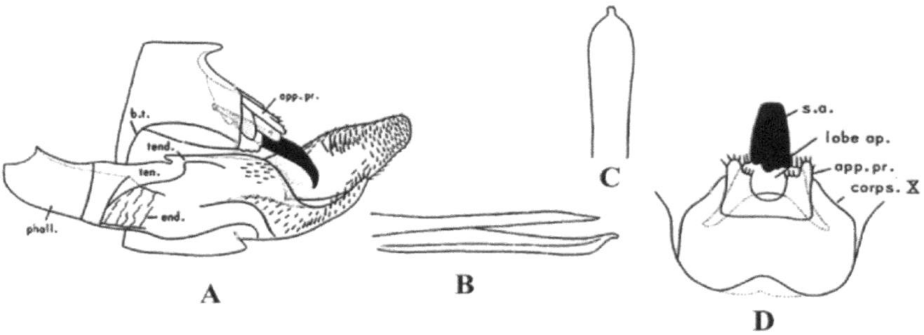

Figure 6.10 *Himalopsyche yatrawalla* (♂): (A) Lateral view of male genitalia; (B) Phallic apparatus; (C) Phallus as seen from above; (D) Dorsal view of Xth segment (from Schmid, 1966).

Genitalia (♀): (Figs. 6.11A–B). The females of *Himalopsyche yatrawalla* are identified based on a sternal plate of the IXth segment which is not very concave.

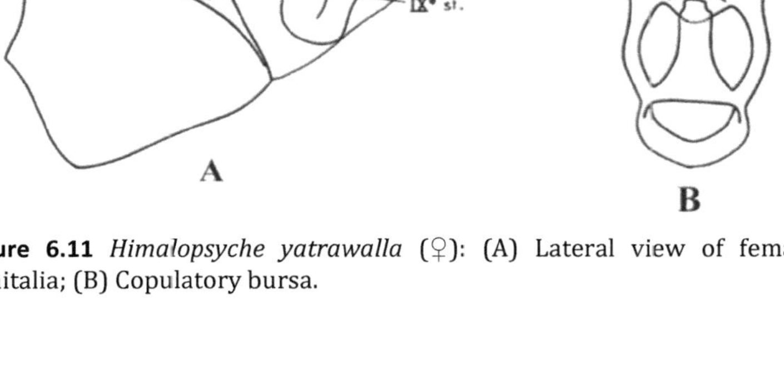

Figure 6.11 *Himalopsyche yatrawalla* (♀): (A) Lateral view of female genitalia; (B) Copulatory bursa.

Holotype: INDIA: Uttarakhand; Pauri Garhwal, Hilang, 7-VI-1958.

Holotype depository: ROM, Canada.

Distribution: India (Uttarakhand).

9. *Himalopsyche yongma* Schmid

Himalopsyche yongma Schmid, 1963: 215–217.

Adult: Upper head and thorax dark brown, with reddish brown areas and fairly abundant fine golden hairiness. Antennae brown ringed red. Face and palps dark brown. Legs thick, armed with strong claws. Abdomen dark brown, paler below and relatively long. Wings narrow and elongate, but are not very large in relation to the body, which gives the insects a slightly branchy appearance. The fore wings are reddish brown, intensely and evenly spotted with light shades covering the approximately equal area. Posterior wings are brownish, relatively dark and slightly spotted along the anterior apical edge.

Genitalia (♂): (Figs. 6.12A–B). IXth segment is regularly and strongly elongated all around, except dorsally where it is narrowed by half to embed the Xth segment. The upper apical angles are rounded and well-prominent. The Xth segment is not very prominent; dorsal lobes are large, fused to each other for half their length and tapering a little before their ends. Anal scleritis in long stick slightly curved in a hook at the apex. Phallobase short and carrying strong "claspers hangers" widened at their end which is notched to receive the strong tendons of the lower appendages. Phallus in the form of a narrow band, well chitinous, pointed at the apex and slightly concave at the top. The upper median internal part of the appendix bears a large area covered with strong and dense bristles, directed forward and down. The internal lobe reached half the length of the outer lobe and tapered gradually to the tapering apex; the internal face has a few conical spines, very short and thick.

Genitalia (♀): Not available.

Holotype: INDIA: Sikkim, Tangshing, 12800 ft., 15-IV-1959.

Holotype depository: ROM, Canada.

Distribution: India (Sikkim).

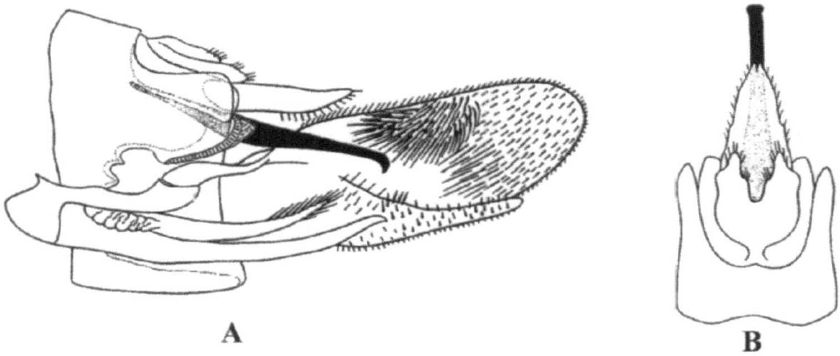

Figure 6.12 *Himalopsyche yongma* (♂): (A) Lateral view of male genitalia; (B) Dorsal view of IXth and Xth segments (from Schmid, 1963).

10. *Himalopsyche amitabha* Schmid

Himalopsyche amitabha Schmid, 1966: 215–217.

Adult: Head and thorax dark brown, with reddish brown areas and abundant fine golden hairs. Antennae brown ringed red. Face and palps dark brown. Legs thick, armed with strong claws. The body and wings are well-colored.

Genitalia (♂): (Figs. 6.13A–B). The dorsal part of the IXth segment is more developed towards the rear. The lobes of the base of the Xth segment are shorter, the lower appendages more massive and the parameres are simple.

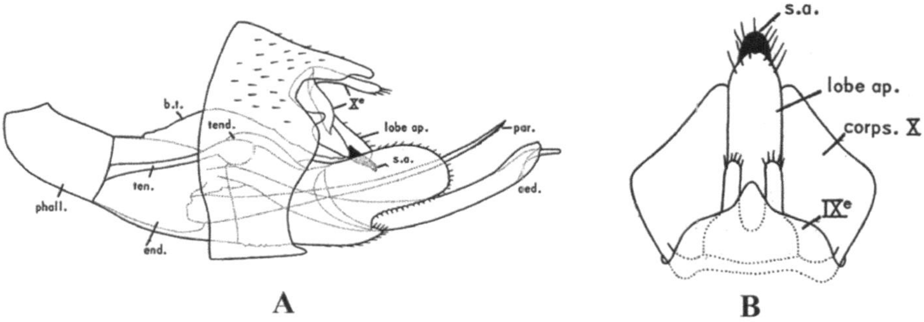

Figure 6.13 *Himalopsyche amitabha* (♂): (A) Lateral view of male genitalia; (B) Dorsal view of Xth segment (from Schmid, 1966).

Genitalia (♀): (Figs. 6.14A–C). The lateral concavity of the VIIIth segment is marked and the vulvar part smaller. Copulatory bursa slightly larger in size than other species of the genus.

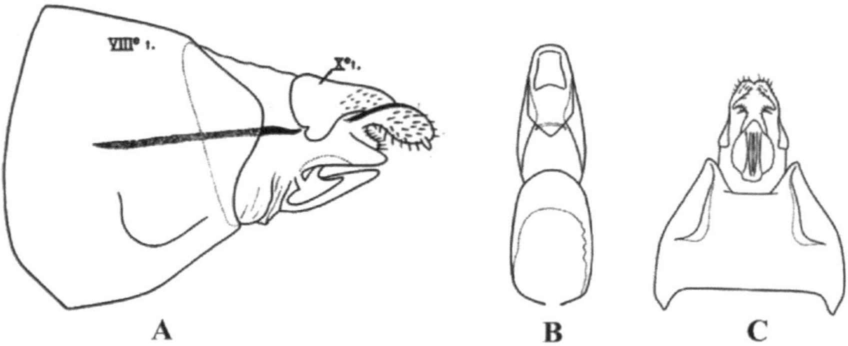

Figure 6.14 *Himalopsyche Amitabha* (♀): (A) Ventral view; (B) Copulatory bursa.

Holotype: Sikkim, Tanggu; 12,000–12,800 ft; 19-VI-1959.

Holotype depository: ROM, Canada.

Distribution: India (Sikkim).

11. *Himalopsyche angnorbui Tangguh* Schmid

Himalopsyche angnorbui Schmid, 1963: 206–208.

Subspecies: *Himalopsyche angnorbui sherpa* Schmid, 1963: 206–208.

Adult: Entire body brown but antenna with alternate dark and light brown bands and forewing with dark brown irrotations. Length of antenna 18.43 mm and forewing 22.8 mm. Wings large, long and narrow; the anterior has a dark brown background staining, lighter along the axis of the cells. In forewing, fork 1st arises prior to 2nd. Hind wings are greyish brown, spotted lightly at the tips.

Genitalia (♀): (Figs. 6.15A–B). Tergum longer than sternum. The anterior end of the VIII segment with triangular projection and the upper end of the posterior margin is convexly produced. Segment IX and X raised upward and apodeme extending anterad up to an almost entire length of VIII segment. Vaginal apparatus apically with sclerotization markings and basally with paired oval structures.

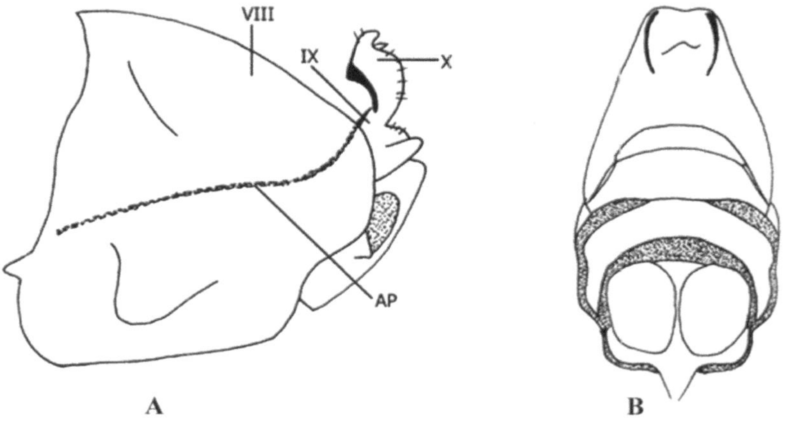

Figure 6.15 *Himalopsyche angnorbui* (♀): (A) Lateral view of female genitalia; (B) Ventral view of vaginal apparatus.

Genitalia (♂): (Figs. 6.16A–E). IXth segment very regularly elongated all around its periphery and with its apical dorsal border slightly extended into a short, rounded tongue, forming a roof. In this species, the Xth segment has a shape and arrangement unique in the genus. It is a plate, which is tilted 180° downwards and forwards and thus constitutes the ceiling of the genital cavity; the basal end of the segment therefore has an apical position and is prolonged by a very short and wide tongue, bearing in its center two long and very slender appendages, in the form of fingers and slightly thickened in buttons at their ends. There is no doubt that these two pieces play the original role of the Xth segment, promoted to other functions by its invagination which is not reversible. There is an interesting case of vicarianism there. The Xth segment consists of a long central part, terminated by two lobes, appearing as claws seen laterally and flanked by two other obtuse lateral lobes; the central part is accompanied by two side lobes reaching their own size and tapering at their ends; the relief of the Xth segment is quite complex, although not very marked. The anal sclerites are present and appear as a small claw. Very large phallic apparatus. Phallobase is cylindrical, longer than high and bearing very long "claspers hangers" widened to their ends. Phallus is moderately chitinous, having a T-section at its base; long, slender and ending in a curve which gives it the shape of an ostrich head. Inferior appendages small and mono-articulate; these are two lobes oblique upwards, rather slender, without thorns and long ovals at their ends. Parameres are thin and very chitinous, formed of three branches of unequal size and bearing small thorns of varying size.

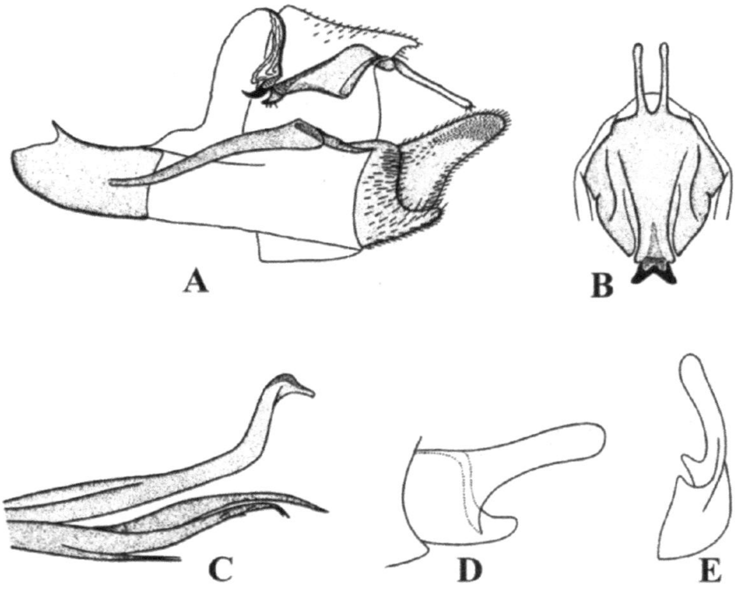

Figure 6.16 *Himalopsyche angnorbui* (♂): (A) Lateral view of male genitalia without phallic apparatus; (B) Dorsal view of Xth segment; (C) Phallic apparatus; (D) Inferior appendage seen ventrally; (E) Inferior appendage of *Himalopsyche angnorbui sherpa* seen laterally (from Schmid, 1963).

Holotype: INDIA: Uttarakhand; Pauri Garhwal, Binaik Chatti, 7,000 ft.; 2-VII-1958.

Holotype depository: ROM, Canada.

Additional material examined: INDIA: Himachal Pradesh: Marhi, 2700 m, 08. viii. 2008. (Coll. Parey & Pandhar).

Distribution: Nepal, India (Himachal Pradesh, Sikkim and Uttarakhand).

12. *Himalopsyche horai* (Martynov)

Rhyacophila pallida Martynov, 1935: 100–102, Fig. 7, ♂ (Name pre-occupied by Mosely, 1930: 149):

"Himachal Pradesh: Punj-Pul Nullah, 2100 m." (ZSI Kolkata, ♂). (Coll. Parey & Pandhar).

Rhyacophila horai Martynov, 1936: 306 (Replacement Name).

Himalopsyche horai (Martynov): Banks, 1940: 197 (New Combination).

= *H. phedongensis* Kimmins, 1952: 356–357 (Synonymized by Schmid, 1966: 152).

Adult: Entire body pale brown except meso and metathorax which are brownish in color. Forewing with dark brown irrotations on radial area. Length of antenna 11.64 mm and forewing 15.52 mm. In the forewing fork, 2nd arises slightly before 1st.

Genitalia (♂): (Figs. 6.17A–C). In dorsal view preanal appendage flap shaped. Segment IX is broader in its dorsal than in ventral parts. Preanal appendage forms a very broad & long plate, apical portion of which is curved downwards. The basal segment of the inferior appendage is large, nearly elliptical and constricted at the base. The terminal segment is very small, minute and curved inwards. Anal sclerite rounded and phallicata curved upwards. In ventral view terminal segment of inferior appendage digitiform.

Genitalia (♀): Not available.

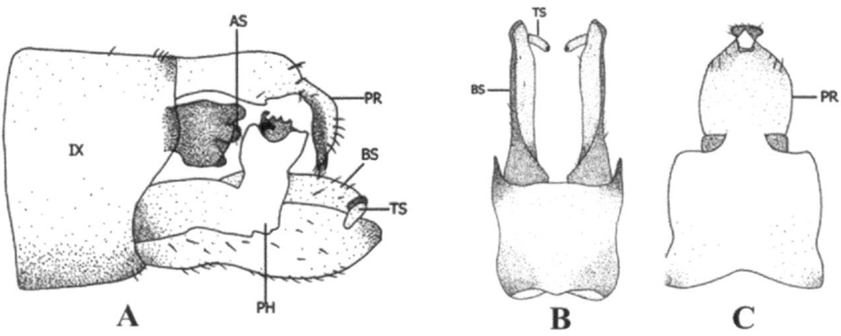

Figure 6.17 *Himalopsyche horai* (♂): (A) Lateral view of male genitalia; (B) Ventral view; (C) Dorsal view.

Additional material examined: INDIA: Arunachal Pradesh: Tato, 1700 m, 28-IV-2010; Uttarakhand: Shyna Chatti, 2000 m, 27-IX-2008; Munsiyari, 2200 m, 20-VI-2011; Hanuman Chatti, 3200 m, 24-VI-2011. (Coll. Parey & Pandhar).

Material depository: ZSI, Kolkata (India).

Distribution: Bhutan, Nepal, India (Arunachal Pradesh, Himachal Pradesh, Sikkim and Uttarakhand).

13. ***Himalopsyche lanceolata* (Morton) *Rhyacophila***

Himalopsyche lanceolata Morton, 1900: 2–3

Adult: Wings long with a semi-curved tip and a radius that is curved apically; apical fork 1 extends inwardly further than fork 2. Costal hairs are simple. Greyish-yellow or stramineous forewings; pale anterior part with a few elongate dots on the veins; coastal area slightly irrorated; a prominent brown dot on the costal border immediately before the apex. Apical cells 2 and 3 are brownish grey, the band so formed being continued obliquely towards the inner peripheral area. The inner marginal area is closely irrorated with dark grey; post-costa is densely pubescent and mostly black as are also the veins converging at the arculus. Hind wings are paler, hyaline and slightly yellowish at the apices. Antennae pale yellow; head pronotum and legs yellowish; meso and meta-nota brownish; abdomen dark fuscous.

Genitalia (♂): (Figs. 6.18A–D). IXth segment not produced dorsally; lateral margin slightly excised at the attachment of inferior appendage. Xth tergal segment forms a hood, bifid to about halfway, the two branches narrowing to slender, down-curved, asymmetrical spines with two or three strong, lanceolate setae at their anterior end. 10th sternal segment membranous. Phallus with a large, semi-membranous basal sheath. Sheaths as long as phallus, incurved, spiniform, with one or two lanceolate setae about midway. Inferior appendage with a large, oblong basal segment, apex slightly concave. The apical segment is very small, quadrate and curved downward, the inner surface is equipped with spinules.

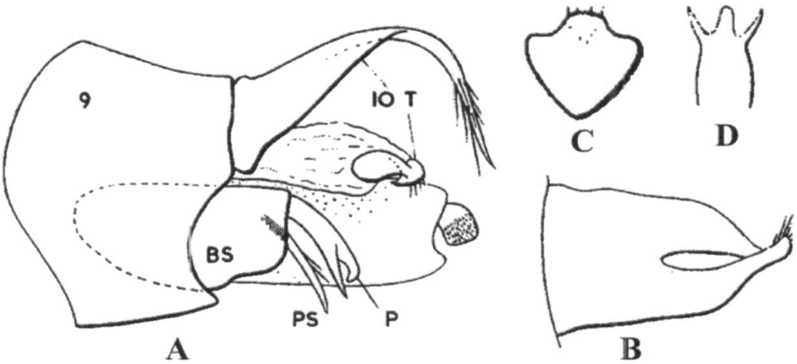

Figure 6.18 *Himalopsyche lanceolata* (♂): (A) Lateral view of male genitalia; (B) Dorsal view of 10th tergite; (C) Sclerotized plate of 10th tergite (dorsal); (D) Apex of penis, ventral (from Kimmins 1952).

Genitalia (♀): (Figs. 6.19A–C). VIIIth segment long, obliquely truncate from the side, deeply notched from above, inflated from below, and truncate apically. The segment is divided about midway by a false suture, deeply emarginate laterally. There is a short, stout, basal apodeme. IXth segment very small, triangular from the side, with long, slender apodemes, and is more or less fused with the Xth segment to form a short hood with rough margins, and carries a pair of microscopic, single-segmented cerci.

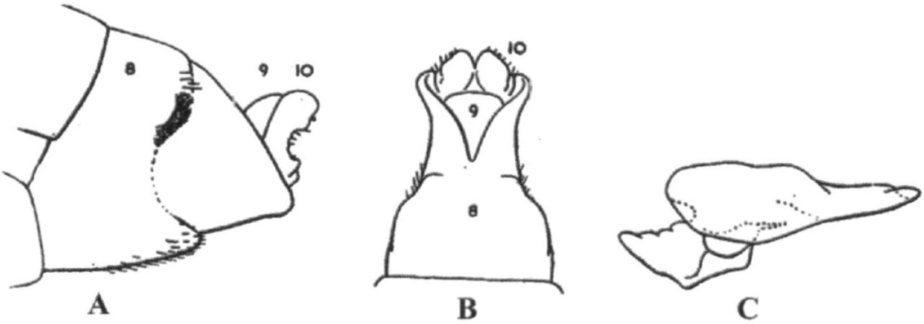

Figure 6.19 *Himalopsyche lanceolata* (♀): (A) Lateral view of the apex of abdomen; (B) Dorsal; (C) Copulatory bursa, lateral (from Kimmins 1952).

Holotype: ASSAM: Khasi Hills.

Holotype depository: ROM, Canada.

Distribution: Myanmar, India (Meghalaya, Manipur).

14. *Himalopsyche lepcha* Schmid

Himalopsyche lepcha Schmid, 1963: 222–223

Adult: Head entirely greenish gray, with the antennae and palps uniformly reddish. Above the thorax reddish yellow, with fine bicolored, reddish and brown hairiness. Weeping and front legs are yellow gray; weeping and middle and hind legs are light reddish yellow. Abdomen blackish above, yellowish below. Wings rather large, broad and with well-rounded ends. The anterior ones have a rather strongly contrasting coloring; their background color is golden yellow, with numerous macules and brown areas; the center of the wing is yellow, with a few small brown spots; the anastomosis is quite strongly outlined in brown; the base of apical cells has a long, uniformly yellow transverse band, followed by another brown band. The base of the wing is brown, but the subthyridial cell is very dark with two very distinct silvery spots. Anal and post-costal cells are uniformly yellow. In **females**, the head is light brown orangish with dark brown pubescence. Antenna pale yellow with black pubescence. Maxillary palp, labial palp, thorax and abdomen pale yellow with brown pubescence. The foreleg is pale yellow but due to the presence of dark brown pubescence beyond femur of foreleg it appears dark brown and pale yellow up to femur, mid and hind legs pale yellow.

Genitalia (♀): (Figs. 6.20A–B). Anterior margin of segment VIII with a broad concavity at lower end and simple in outline. Posterior margin of segment VIII half in length as compare to the anterior margin. Apodeme slightly extending anterad to segment VII. In ventral view base of vaginal apparatus H-shaped.

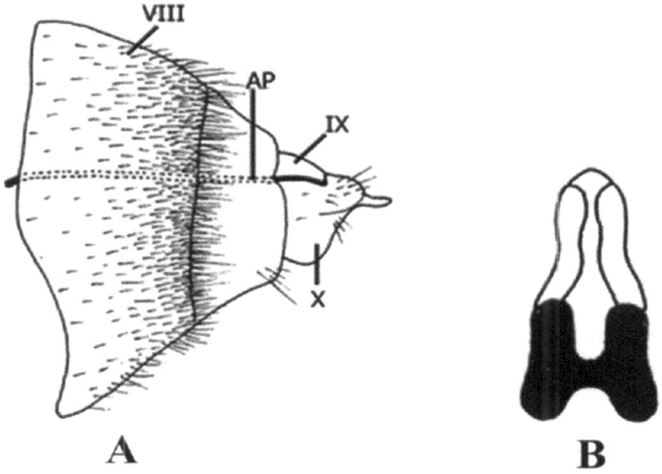

Figure 6.20 *Himalopsyche lepcha* (♀): (A) Lateral view of female genitalia; (B) Ventral view of vaginal apparatus.

Genitalia (♂): (Figs. 6.21A–C). IXth segment not very elongated, with a strong and wide indentation at the level of the lower appendages and a well-developed lateral angle; dorsally, it constitutes an appendage three times as long as it is wide and rounded at its end; this is concave at its lower face and seems to be made up of the IXth segment alone, to the exclusion of the Xth segment. The latter is large, massive, twice as long as it is high, invaginated in the IXth segment at its base and made up of two large lateral lobes. Anal scleritis is very small, fully concealed and a little chitinous; seen from above, it appears subcircular and indented over half its length. Phallobase is small; with "claspers hangers", but a thin chitinous band connects the phallobase to the lower appendages. Large phallotheca. The aphallus is a subcylindrical organ, well chitinous, curved in S upwards and the extreme base of which has the section of a T. Paramers twice shorter than the phallus, in strong cylindrical appendages, pointed and a little thickened before their ends. Lower appendages are clearly biarticulate. Wingspan 24 mm.

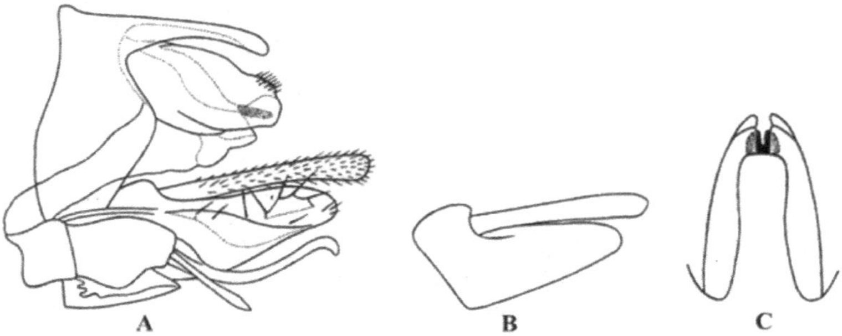

Figure 6.21 *Himalopsyche Lepcha* (♂): (A) Lateral view of male genitalia; (B) Lower appendage seen from the side; (C) Dorsal appendix of the IXth and Xth segment seen from above (from Kimmins 1952).

Holotype: Western Bengal, Bara Hata (5,000 ft.) 22-III-1959.

Holotype depository: ROM, Canada.

Additional material examined (♀): INDIA: Sikkim: Lachen, 3200 m, 24-V-2011. (Coll. Parey & Pandhar).

Distribution: Bhutan, India (Sikkim and West Bengal).

15. *Himalopsyche biansata* Kimmins

Himalopsyche biansata Kimmins, 1952: 351–352

Adult: Yellowish head with dark brown warts. Antennae brownish with pale yellow annulations; basal joint yellow; palpi harder. Thorax brown, the median stripe of mesonotum paler and testaceous. Legs spiny, yellowish. Anterior wings are broad, with regular fuscous reticulation. Fork 3 in both wings is sessile and fork 5 in the forewings is rather broader. Abdomen brownish dorsally, yellowish ventrally; IXth and Xth segments yellowish.

Genitalia (♂): (Figs. 6.22A–B). IXth segment with dorsal terminus widely notched. Lateral margin with a wide V-shaped excision in the lower half. The outer plate does not foliate, rather broad at its base in lateral view, lower margin concave in the apical half so that the two outer plates resemble the handles of a wheelbarrow. Xth segment membranous. Phallus long, fairly stout, at the base with downwardly directed flanges on each side, then dilating before the apex. There is a single, spiniform sheath, nearly as long as the phallus. Inferior appendage two-segmented, basal segment long, broad at the base. The terminal segment short widened apically.

Genitalia (♀): Not available.

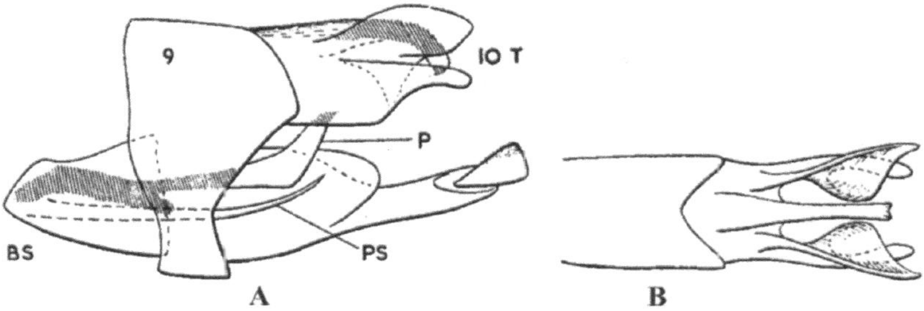

Figure 6.22 *Himalopsyche biansata* (♂): (A) Male genitalia, lateral; (B) Dorsal view of IXth and Xth tergites (from Kimmins, 1952).

Holotype: TIBET Yatung, 4500 ft.

Holotype depository: ROM, Canada.

Distribution: Tibet, Bhutan, India (Sikkim).

16. *Himalopsyche tibetana* (Martynov) *Rhyacophila*

Rhyacophila tibetana Martynov, 1930: 65–67

Himalopsyche tibetana (Martynov): Banks, 1940: 197

Adult: In general appearance, this species resembles *H. biansata* but rather smales and is less pale than the male. Nervures yellowish testaceous; subcosta with an indistinct fork; first apical fork beginning a little earlier than the second. Posterior wings (♂) are greyish and somewhat brownish or fumose; fork 2 begins somewhat earlier than fork 1. In the posterior wings (♀) fork 2 begins considerably earlier than fork 1; in the anterior wings both forks begin approximately at the same level. Abdomen brownish above, yellowish ochraceous beneath; ninth and tenth segments yellowish (♂) (Martynov, 1930).

Genitalia (♂): (Figs. 6.23A–C). IXth segment elevated apically in a narrow ridge. Xth segment is lightly sclerotized, complex, partly fused to the IXth. Each inner plate is directed upwards and is shorter and blunter than the outer, which is tapering at the apex. The phallus is slender, long and curved upwards dorsally with a dilation before the apex. Inferior appendages are two-segmented, basal segment large, slender, terminal segment small, slender, apex slightly excised and spinulose.

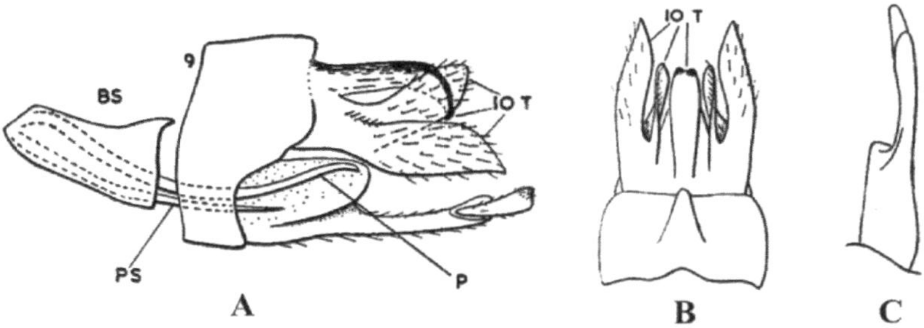

Figure 6.23 *Himalopsyche tibetana* (♂): (A) Male genitalia, lateral view; (B) Dorsal view of IXth and Xth segments; (C) Ventral view of inferior appendages (from Kimmins, 1952).

Genitalia (♀): (Figs. 6.24A–C). Dorsally, VIII segment near its posterior margin with slight concavity. Sternum of segment VIII broader than tergum. Segment VIII roughly rectangular in shape. Segment IX outlined by heavily sclerotized apodeme. Segment X rounded apically and with terminal papilla which is directed downwards. Apodeme extending anterad up to half the length of VIII segment. VIII segment with ventral lobe which is rounded apically and shorter than segment X. In ventral view, ventral lobe divided up to half of its total length and segment VIII roundly produced. Vaginal apparatus slipper shaped.

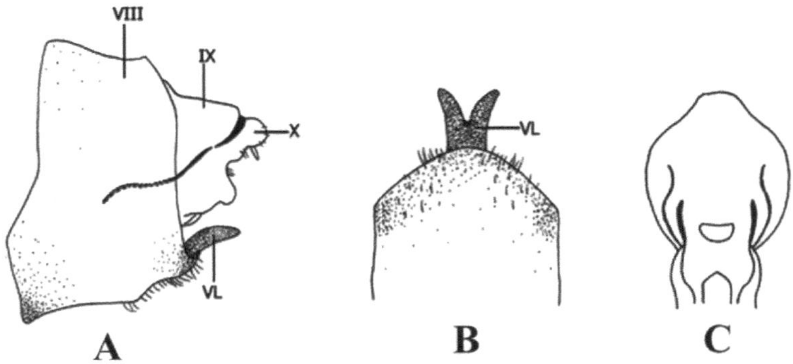

Figure 6.24 *Himalopsyche tibetana* (♀): (A) Lateral view of female genitalia; (B) Ventral view of ventral lobe; (C) Ventral view of vaginal apparatus.

Holotype: TIBET: Yatung, 4500 ft., A. E. Hobson. Chumbi, 11,000 ft., 20. vi., 21. vii. 1924, R.W.G. Hingston.

Holotype depository: ROM, Canada.

Additional material examined: INDIA: Jammu & Kashmir: Sonmarg, 3000 m, 26.vii.2011, 1 ♀; Himachal Pradesh: Losar, 3200 m, 22. viii. 2008, 2 ♂♂; Chhatradu, 3600 m, 21.viii.2008, 6♂♂, 1♀; Marhi, 2700 m, 08.viii.2008, 2♂♂; Kaza, 3000 m, 24.viii.2008, 1♂, 1♀; Uttarakhand: Badrinath, 3400 m, 22.vi.2012, 1♀. (Coll. Parey & Pandhar).

Distribution: Bhutan, Nepal, Tibet, India (Sikkim, Uttarakhand, Jammu & Kashmir and Himachal Pradesh).

17. *Himalopsyche digitata* (Martynov) *Rhyacophila*

Rhyacophila digitata Martynov, 1935: 102–103

Himalopsyche digitata (Martynov): Banks, 1940: 197

Adult: Head, antenna, maxillary palp, prothorax, metathorax brown; labial palp pale brown; mesothorax and tibia of the foreleg and midleg dark brown; forewing with dark brown irrotations. Length of antenna 15.2 mm and forewing 21.46 mm. In forewing, fork 1st arises prior to 2nd.

Genitalia (♂): (Figs. 6.25A–C). In the dorsal view, segment X with pointed lobes. In the lateral view, the tergum of segment IX is convexly raised. Anterior margin of segment IX without concavity and posterior margin with concavity near the lower end. Tergum of segment IX longer than sternum. The preanal appendage is divided into two parts, the upper part being longer and broader than its lower one. Segment X curved downwards. The base of phallicata with sinuous outline and is apically curved upwards. Terminal segment of inferior appendage with prominent black spine. In ventral view apex of the inferior appendage is rounded.

Genitalia (♀): (Figs. 6.26A–C). Segment VIII with deep concavity and anterior margin of segment VIII raised upwards. Segment IX roundly produced and apodeme extending anterad beyond segment VIII. Segment VIII with ventral lobe which is cylindrical in shape. Vaginal apparatus with triangular projection at the centre apically.

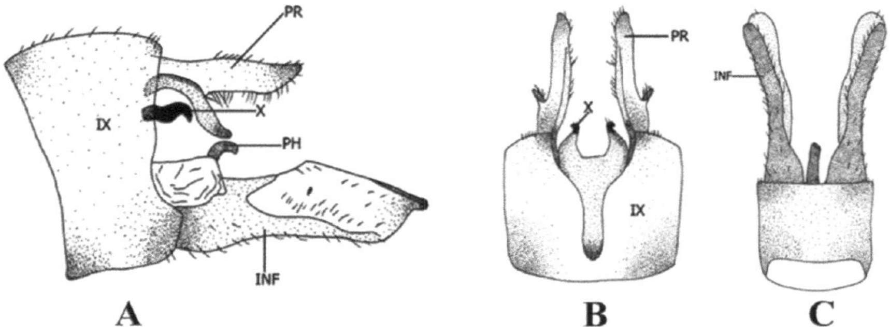

Figure 6.25 *Himalopsyche digitata* (♂): (A) Lateral view of male genitalia; (B) Dorsal view; (C) Ventral view.

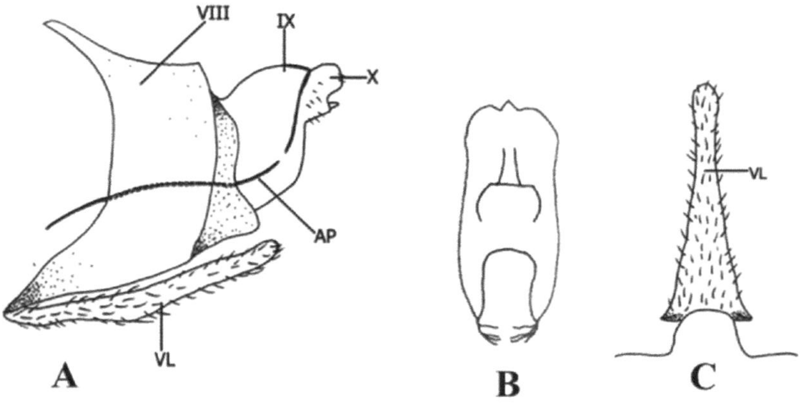

Figure 6.26 *Himalopsyche digitata* (♀): (A) Lateral view of female genitalia; (B) Ventral view of vaginal apparatus; (C) Ventral view of ventral lobe (from Schmid, 1966).

Additional material examined: INDIA: Uttarakhand: Pingla Pani, 1300 m, 11.vi.2011; Chopta, 2700 m, 13.vi.2009; Joshimath, 2000 m, 18.vi.2011; Arunachal Pradesh: Lumla, 2300 m, 15.v.2011; Bomdila, 2400 m, 03.x.2010; Sikkim: Lachen, 3200 m, 24.v.2011. (Coll. Parey & Pandhar).

Material depository: ZSI, Kolkata.

Distribution: Nepal, Bhutan, India (Sikkim, Arunachal Pradesh, Uttarakhand, West Bengal).

18. *Himalopsyche hierophylax* Schmid

Rhyacophila hierophylax Schmid, 1966: 150–151

Adult: Body and wing coloration similar to *H. digitata*, although dark oblique areas of forewings are less visible or absent. The face of insect, however, is quite different. In *hierophylax*, the abdomen is strongly telescoped and the wings very reduced; the antennae and the legs have a normal length, that they appear very long. The forewings protrude only slightly from the end of the abdomen in smaller insects and reach twice the body length in larger ones, but the ♀ consistently has smaller wings than the ♂.

Genitalia (♂): (Figs. 6.27A–B). Genitalia of male is similar to those of *H. digitata*, but the latter are much less marked, especially as regards the preanal appendages.

Genitalia (♀): Not available.

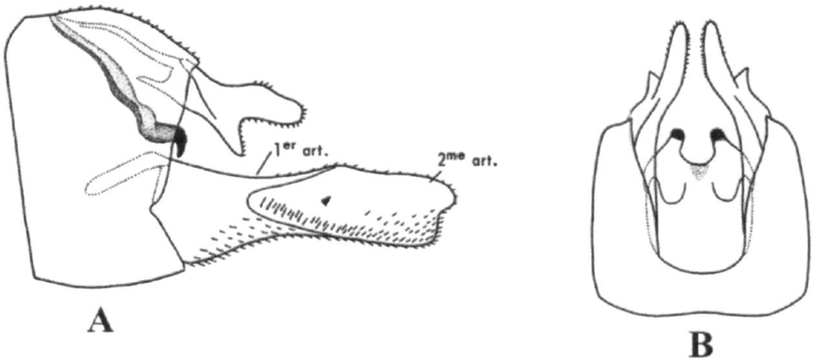

Figure 6.27 *Himalopsyche hierophylax* (♂): (A) Male genitalia, lateral view; (B) Dorsal view (from Schmid, 1966).

Holotype: India (Pauri Garhwal); Kedarnath, 13-15-V-1958.

Holotype depository: ROM, Canada.

Distribution: India (Uttarakhand).

19. *Himalopsyche maitreya* Schmid

Rhyacophila maitreya Schmid, 1963: 220-221

Adult: Top of head and thorax reddish brown, with paler areas, yellow, reddish, silvery reflections and scant two-tone hair, black and silver, Antennae brown, ringed with yellow. Reddish brown weeping. Legs yellowish, fairly pale, with dark brown rings on the fore and median legs. Abdomen gray-brown above, yellowish-brown below. Fore wings are small compared to the body; they are light brown yellow and intensely spotted with dark gray-brown spots. In forewing forks, 1st and 2nd arise at the same level. Hind wings are greyish yellow, with a few small marginal macules in the apical part of the wing. Nervulation without peculiarities.

Genitalia (♂): (Figs. 6.28A–B). IXth segment regularly elongated all around, but deeply indented dorsally for the insertion of the Xth segment. The latter is a downward slanting plate and strongly chitinous; laterally, it appears regularly thin, but thickened at its end which ends in a small even upper hook; Dorsally, the Xth segment appears as a rectangular plate, twice as long as it is wide, with sinuous lateral edges and terminated in three lobes: two lateral wide and obliquely truncated at their end, curved in a hook towards the bottom, rather distant one from the other and between which is a 3rd lobe, less chitinous and located on a lower plane than the previous ones. This 3rd lobe probably represents reduced anal scleritis. The phallic apparatus shows well developed "claspers hangers", quite chitinous, in the form of a subvertical band and provided with a posterior plate on which the tendon of the corresponding inferior appendage is inserted. Phallobase short, thick and membranous. Phallus with a basal plate located under that of the phallobase and appearing, over the rest of its length, like a thin cylinder, very chitinous, sinuous and curved in a hook at its end. Parameres absent. The phallotheca is highly developed, as long as the phallus which it surrounds like a sheath laterally and ventrally. Bi-articulated inferior appendages; the basal segment is thin and tall; it is pointed at its end and indented at its lower apical part.

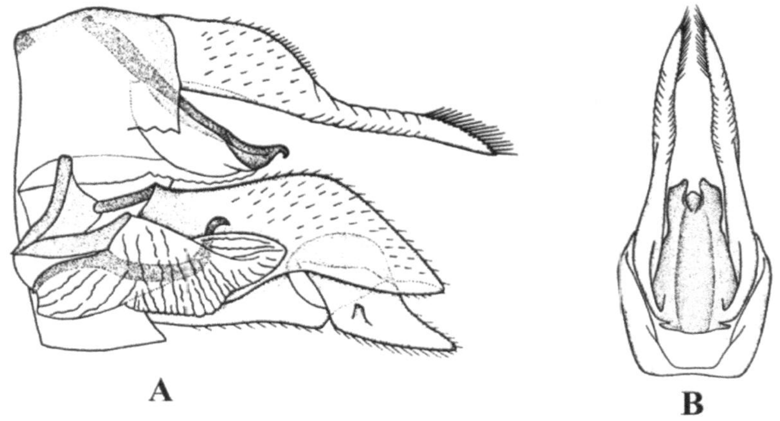

Figure 6.28 *Himalopsyche maitreya* (♂): (A) Lateral view of male genitalia; (B) Dorsal view of IXth and Xth segments (from Schmid, 1963).

Genitalia (♀): (Figs. 6.29A–C). The postero-dorsal margin of segment VIII roundly produced and with a slight concavity prior to its tip. Ventrally segment VIII gives rise to a prominent finger-like ventral lobe. Apodeme extending anterad to segment VII. The apex of VIII sternite heart-shaped. Vaginal apparatus with U shape indentation apically.

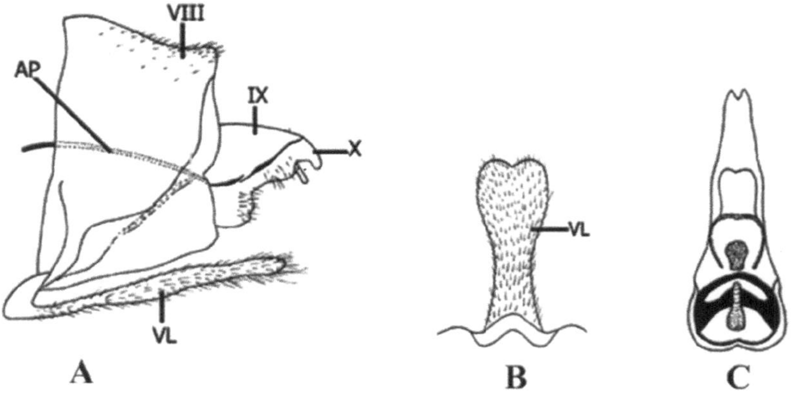

Figure 6.29 *Himalopsyche maitreya* (♀): (A) Lateral view of female genitalia; (B) Ventral view of ventral lobe; (C) Ventral view of vaginal apparatus.

Holotype: Pauri Garhwal, Akrotkoti (5,700 ft.) 17-18-V-1958.

Holotype depository: ROM, Canada.

Additional material examined: INDIA: Arunachal Pradesh: Lumla, 2300 m, 07-X-2010. (Coll. Parey & Pandhar).

Distribution: Nepal, India (Arunachal Pradesh and Uttarakhand).

Chapter 7

Genus *Rhyacophila* Pictet, 1834

Rhyacophila Pictet, 1834: 181.

Type species: *Rhyacophila vulgaris* Pictet, 1834 (designated by Ross, 1944: 32)

Ross, 1956 & Schmid, 1970 (genus revised)

Diagnostic features: Spurs 3,4,4. The first and second segments of maxillary palp are short, the second almost subglobular, the third very big, the fourth and fifth each not quite as long as the third, and the fifth terminating in a minute pointed apex (Fig. 2.1B). Metascutellum without spiniferous warts. Forewing with apical forks I–V present, hindwing with forks I–III & V present, discoidal cell open in both wings; the branches of R are anterior to the wing tip and those of M are posterior to it (Fig. 2.2A). Legs long and slender, with few spines, spurs long and distinct. Abdomen with a small callosity (Fig. 2.1C) and carina on each side of the sternite V. In the male there may be a ventral process on the sternite VII, and sometimes also on the VI but in female on VI only; inferior appendage always two-segmented.

Branches of Genus *Rhyacophila* Pictet, 1834:

(I) *R. vulgaris* Branch
(II) *R. divaricata* Branch
(III) *R. philopotamoides* Branch
(IV) *R. naviculata* Branch

Rhyacophilidae of India
Sajad Hussain Parey, Tabraq Ali, and Manpreet Singh Pandher

ISBN 978-981-4968-74-4 (Hardcover), 978-1-003-41095-9 (eBook)
www.jennystanford.com

I. Diagnostic features of *Rhyacophila vulgaris* Branch

- Segment IX with dorsal apical lobe.
- Segment X not prominent and vertical in orientation.
- Preanal appendage free.
- Anal sclerite long.
- Apical band long, well sclerotized, and quite prominent.
- Tergal band sclerotized.
- Phallic apparatus complete.

Schmid (1970) divided this branch into nine Subgroups, but Indian species are included in only one Subgroup, the *Rhyacophila anatina* Subgroup.

Indian species of *Rhyacophila anatina* Subgroup:

1. *R. anatina* Morton, 1900
2. *R. bhotia* Schmid, 1970
3. *R. dongkyapa* Schmid, 1970
4. *R. drotangpa* Schmid, 1970
5. *R. himachalensis* Kaur & Saini, 2012
6. *R. kadampa* Schmid, 1970
7. *R. lhadzongpa* Schmid, 1970
8. *R. manipuri* Schmid, 1970
9. *R. nagongpa* Schmid, 1970
10. *R. yipung* Schmid, 1970
11. *R. yigrongpa* Schmid, 1970

II. Diagnostic features of *Rhyacophila divaricata* Branch

- Preanal appendage long.
- Phallic apparatus long and complete.
- Apical band present only in *R. nigrocephala* Group.
- Tergal band sclerotized or membranous.

This branch is divided into four Groups, but Indian species fall under three Groups viz:

(a) *R. nigrocephala* Group

(b) *R. divaricata* Group

(c) *R. curvata* Group

(a) *R. nigrocephala* Group (Characteristics)

- Dorsal apical lobe of IXth segment long.
- Preanal appendage long and voluminous.
- Phallic apparatus simple.

Subgroups of *R. nigrocephala* Group: *R. nigrocephala* Group is represented by two Subgroups:

(1) *R. nigrocephala* Subgroup

(2) *R. hingstoni* Subgroup

Key to Subgroups of *R. nigrocephala* Group

- Paramere absent.................................***R. nigrocephala* Subgroup**
- Paramere present.......................................***R. hingstoni* Subgroup**

(1) *R. nigrocephala* Subgroup: This Subgroup is represented by 18 species in India and its surrounding areas and is probably the largest one from the angle of species diversity.

1. *R. kawachenpa* Schmid, 1970
2. *R. kangjongpa* Schmid, 1970
3. *R. nyukmadongpa* Schmid, 1970
4. *R. scotina* Kimmins, 1953
5. *R. drosampa* Schmid, 1970
6. *R. tecta* Morton, 1900
7. *R. chakungpa* Schmid, 1970
8. *R. putata* Kimmins, 1953
9. *R. lhabu* Schmid, 1970
10. *R. sikungpa* Schmid, 1970
11. *R. nakpo* Schmid, 1970
12. *R. sumdopa* Schmid, 1970
13. *R. tsiudmarpo* Schmid, 1970
14. *R. polha* Schmid, 1970
15. *R. chomoyuma* Schmid, 1970
16. *R. langdarma* Schmid, 1970
17. *R. senggepa* Schmid, 1970
18. *R. monyulpa* Schmid, 1970

(2) ***R. hingstoni* Subgroup:** This Subgroup covers only seven species from India and its neighboring countries such as Bhutan, Tibet (Southern China), etc.

1. *R. khiympa* Schmid, 1970
2. *R. ngulpa* Schmid, 1970
3. *R. hingstoni* Martynov, 1930
4. *R. kedara* Schmid, 1970
5. *R. kanichka* Schmid, 1959
6. *R. hobsoni* Martynov, 1930
7. *R. kadaphes* Schmid, 1959

(b) ***R. divaricata* Group (Characteristics)**

- Preanal appendage long.
- Apical band absent and tergal band membranous.
- Phallicata three-branched.

Subgroups of *R. divaricata* Group: The following 16 Subgroups are included in this Group:

1. **Indian species of *R. nyelungpa* Subgroup:** *R. nyelungpa* Schmid, 1970.
2. **Indian species of *R. dafla* Subgroup:** *R. dafla* Schmid, 1970.
3. **Indian species of *R. khimbarpa* Subgroup:**
    1. *R. rhombica* Martynov, 1935
    2. *R. gyelbu* Schmid, 1970
    3. *R. chembo* Schmid, 1970
    4. *R. muktepa* Schmid, 1970
    5. *R. aureomaculata* Schmid, 1970
    6. *R. aureostigma* Schmid, 1970
    7. *R. choprai* Martynov, 1935
    8. *R. angden* Schmid, 1970
    9. *R. khimbarpa* Schmid, 1970
    10. *R. chumikpa* Schmid, 1970
    11. *R. khasiorum* Schmid, 1970
    12. *R. dakshi* Schmid, 1970
    13. *R. paurava* Schmid, 1959
4. **Indian species of *R. tengyelingpa* Subgroup:** *R. tengyelingpa* Schmid, 1970.

5. **Indian species of *R. chematangpa* Subgroup:**
    1. *R. chematangpa* Schmid, 1970
    2. *R. chimdro* Schmid, 1970
    3. *R. chugalungpa* Schmid, 1970
    4. *R. kyungpa* Schmid, 1970
6. **Indian species of *R. inconspicua* Subgroup:** *R. inconspicua* Morton, 1900.
7. **Indian species of *R. nigrorosea* Subgroup:**
    1. *R. tungkorpa* Schmid, 1970
    2. *R. similis* Martynov, 1935
    3. *R. lhakpa* Schmid, 1970
    4. *R. nigrorosea* Schmid, 1959
8. **Indian species of *R. parva* Subgroup:** *R. parva* Kimmins, 1953.
9. **Indian species of *R. kubra* Subgroup:** *R. kubra* Schmid, 1970.
10. **Indian species of *R. yishepa* Subgroup:**
    1. *R. yishepa* Schmid, 1970
    2. *R. churongpa* Schmid, 1970
    3. *R. bicolor* Kimmins, 1953
    4. *R. kyimdongpa* Schmid, 1970
    5. *R. marpa* Schmid, 1970
11. **Indian species of *R. chamolungpa* Subgroup:** *R. chamolungpa* Schmid, 1970.
12. **Indian species of *R. nabochenpa* Subgroup:**
    1. *R. shingripa* Schmid, 1970
    2. *R. kashmirensis* Kaur and Saini, 2013
    3. *R. nabochenpa* Schmid, 1970
    4. *R. tungpa* Schmid, 1970
13. **Indian species of *R. divaricate* Subgroup:** *R. assimilis* Kimmins, 1953.
14. **Indian species of *R. yarlungpa* Subgroup:** *R. yarlungpa* Schmid, 1970.

15. **Indian species of *R. gyamo* Subgroup:** *R. gyamo* Schmid, 1970.

16. **Indian species of *R. ancestralis* Subgroup:**

    1. *R. dolingpa* Schmid, 1970.
    2. *R. tolungpa* Schmid, 1970.
    3. *R. ancestralis* Martynov, 1935.

(c) *Rhyacophila curvata* Group (Characteristics)

- IX segment short and thick.
- Preanal appendage long.
- Anal sclerite long.
- Apical band absent and tergal band sclerotized.

Subgroups of *R. curvata* Group: This Group comprises **seven Subgroups** from Indian continent and its surrounding countries:

1. **Indian species of *R. chayulpa* Subgroup:**

    1. *R. chayulpa* Schmid, 1970
    2. *R. dongre* Schmid, 1970
    3. *R. tsona* Schmid, 1970

2. **Indian species of *R. tashidingpa* Subgroup:** *R. tashidingpa* Schmid, 1970.

3. **Indian species of *R. tarkiya* Subgroup:**

    1. *R. yullha* Schmid, 1970
    2. *R. chulukpa* Schmid, 1970
    3. *R. netongpa* Schmid, 1970
    4. *R. milarepa* Schmid, 1970
    5. *R. tarkiya* Schmid, 1970
    6. *R. nyamangpa* Schmid, 1970
    7. *R. lepcha* Schmid, 1970

4. **Indian species of *R. curvata* Subgroup:**

    1. *R. shakangpa* Schmid, 1970
    2. *R. kagyupa* Schmid, 1970
    3. *R. lhopa* Schmid, 1970
    4. *R. mortoni* Kimmins, 1953
    5. *R. narayani* Schmid, 1970
    6. *R. khamakhya* Schmid, 1970

7. *R. zhungpa* Schmid, 1970
8. *R. curvata* Morton, 1900
9. *R. dikkaravasini* Schmid, 1970
10. *R. jayadurga* Schmid, 1970

5. **Indian species of *R. mishmica* Subgroup:**

    1. *R. sanglungpa* Schmid, 1970
    2. *R. trulungpa* Schmid, 1970
    3. *R. mishmica* Kimmins, 1953
    4. *R. lonpo* Schmid, 1970
    5. *R. gyaspa* Schmid, 1970
    6. *R. dirangpa* Schmid, 1970
    7. *R. yonggyapa* Schmid, 1970
    8. *R. laptsapa* Schmid, 1970

6. **Indian species of *R. fletcheri* Subgroup:** *R. fletcheri* (Kimmins, 1952).

7. **Indian species of *R. fletcheri* Subgroup:**

    1. *R. karpa* Schmid, 1970
    2. *R. nyerongpa* Schmid, 1970
    3. *R. ugyenpa* Schmid, 1970

III. Diagnostic features of *Rhyacophila philopotamoids* Branch

- X segment not very tall and preanal appendage absent.
- Apical band small, strongly sclerotized and forms three parts.
- Tergal band well sclerotized.
- Phallic apparatus complete and voluminous but not too long.

R. philopotamoids Branch is divided into three Groups but Indian species come under only one Group *i.e., R. castanea* Group.

Characteristic features of *R. castanea* Group

- X segment small, positioned vertically or horizontally with its prominent dorsal apical angle.
- Anal sclerite small.
- Apical band small with its thin appendage.
- Tergal band long.
- Paramere with voluminous head.

Subgroups of *R. castanea* Group:

There are six Subgroups of *R. castanea*, which include:

1. *R. annulicornis* Subgroup
2. *R. wangpo* Subgroup
3. *R. scissa* Subgroup
4. *R. castanea* Subgroup
5. *R. obscura* Subgroup
6. *R. kashongpa* Subgroup

1. **Indian species of *R. annulicornis* Subgroup:** In *R. annulicornis* Subgroup of *R. castanea* Group, eight species are recorded from Indian faunistic limits. Excepting *R. dilata* all other species are contributed by Schmid, 1970.

    1. *R. nyerpa* Schmid, 1970
    2. *R. sherpa* Schmid, 1970
    3. *R. chandzo* Schmid, 1970
    4. *R. kunma* Schmid, 1970
    5. *R. naga* Schmid, 1970
    6. *R. manlungpa* Schmid, 1970
    7. *R. dilata* Martynov, 1935
    8. *R. tshiringpa* Schmid, 1970

2. **Indian species of *R. wangpo* Subgroup:**

    1. *R. armitageorum* Saini & Kaur, 2012
    2. *R. tashepa* Schmid, 1970
    3. *R. wangpo* Schmid, 1970

3. **Indian species of *R. scissa* Subgroup:**

    1. *R. scissa* Morton, 1900
    2. *R. scissoides* Kimmins, 1953

4. **Indian species of *R. castanea* Subgroup:** This Subgroup includes 11 Indian species and is the largest Subgroup of *R. castanea* Group:

    1. *R. namgyal* Schmid, 1970
    2. *R. lobsang* Schmid, 1970
    3. *R. dorje* Schmid, 1970
    4. *R. tsering* Schmid, 1970
    5. *R. norbu* Schmid, 1970

6. *R. ngawang* Schmid, 1970
7. *R. acutis* Kaur & Saini, 2012
8. *R. kusang* Schmid, 1970
9. *R. gyaldzen* Schmid, 1970
10. *R. antiope* Malicky, 1997
11. *R. pemba* Schmid, 1970

5. **Indian species of *R. obscura* Subgroup:** From the Indian subcontinent five species of this Subgroup are recorded:
    1. *R. obscura* Martynov, 1927
    2. *R. chenmo* Schmid, 1970
    3. *R. sakyapa* Schmid, 1970
    4. *R. chungse* Schmid, 1970
    5. *R. bidens* Kimmins, 1953
6. **Indian species of *R. kashongpa* Subgroup:** *R. kashongpa* Schmid, 1970.

IV. Diagnostic features of *Rhyacophila naviculata* Branch

The followings are the characteristics of this Branch

- X segment simple and long.
- Preanal appendage absent.
- Anal sclerite small.
- Apical band simple and tergal band strongly sclerotized.
- Phallic apparatus simple.
- Paramere spiniformes.

R. naviculata Branch is composed of **five Subgroups** but the Indian species pertain to only four Subgroups. These include:

1. *R. naviculata*
2. *R. angulata*
3. *R. lieftincki*
4. *R. bifida*

1. **Indian species of *R. naviculata* Subgroup**: This Subgroup is richly distributed in India and its surrounding areas and is mostly confined to the high-altitude regions. All 16 species are recorded under this Subgroup from India and its surrounding areas:

1. *R. extensa* Martynov, 1927
2. *R. stenostyla* Martynov, 1930
3. *R. ngorpa* Schmid, 1970
4. *R. sajadi* Saini & Kaur, 2012
5. *R. kyadongpa* Schmid, 1970
6. *R. procliva* Kimmins, 1953
7. *R. gelukpa* Schmid, 1970
8. *R. rongpa* Schmid, 1970
9. *R. alticola* Kimmins, 1953
10. *R. poba* Schmid, 1970
11. *R. changpa* Schmid, 1970
12. *R. tsongkhapa* Schmid, 1970
13. *R. spinalis* Martynov, 1930
14. *R. naviculata* Morton, 1900
15. *R. dgaldanpa* Schmid, 1970
16. *R. drokpa* Schmid, 1970

2. **Indian species of *R. angulata* Subgroup:** *R. sherchokpa* Schmid, 1970.

3. **Indian species of *R. lieftincki* Subgroup:** *R. kando* Schmid, 1970.

4. **Indian species of *R. bifida* Subgroup:**
    1. *R. trashipa* Schmid, 1970
    2. *R. bhuchanadhara* Schmid, 1970
    3. *R. tshogpa* Schmid, 1970
    4. *R. maitripa* Schmid, 1970

Description of Indian species of *Rhyacophila* Pictet, 1834

1. ***Rhyacophila acutis*** Kaur & Saini

Rhyacophila acutis Kaur & Saini, 2012: 214

Adult: Head, thorax, abdomen and wings dark brown but maxillary palp, labial palp and legs pale brown. Length of antenna 5.82 mm and forewing 8.73 mm. In forewing forks 1st and 2nd arise at the same level.

Genitalia (♂): (Figs. 7.1A–E). In dorsal view segment X with U shape indentation. Segment X in lateral view with rounded projection. Dorsal appendage of phallic apparatus pointed apically. Terminal segment of inferior appendage very small in size and with two patches of spines. Upper process of ventral lobe of phallicata curved backward. In ventral view inner and outer surfaces of terminal segment of inferior appendage are concave and convex respectively.

Genitalia (♀): Not available.

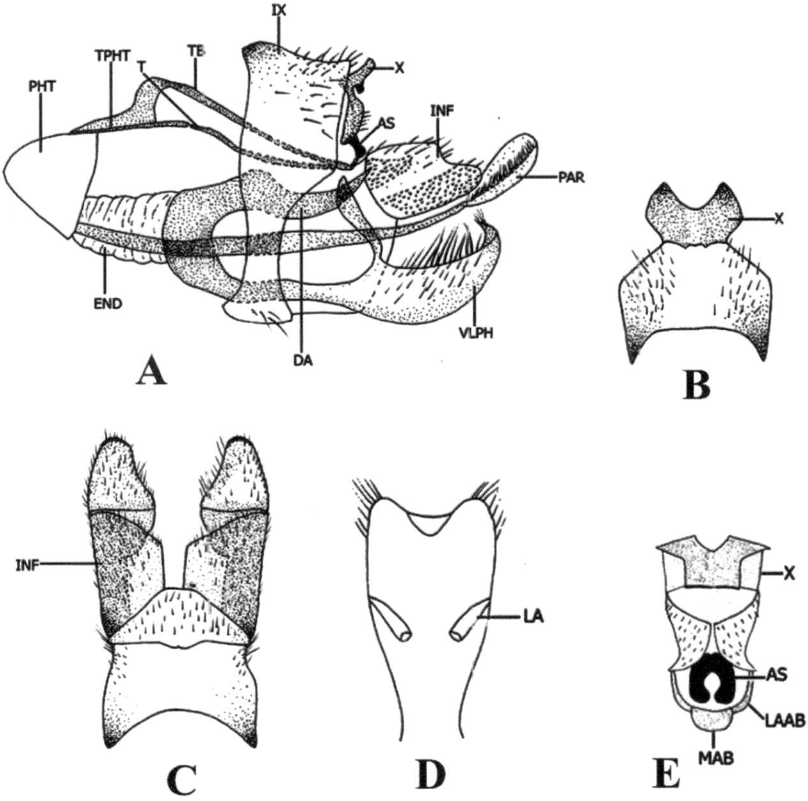

Figure 7.1 *Rhyacophila acutis* (♂): (A) Lateral view of male genitalia; (B) Dorsal view; (C) Ventral view; (D) Dorsal view of ventral lobe; (E) Caudal view.

Holotype: INDIA: Meghalaya, Padammaphlong, 1500 m, 26.v.2011. (Coll. Parey & Pandhar).

Holotype depository: MDZPUP, India.

Distribution: India (Meghalaya).

2. *Rhyacophila alticola* Kimmins

Rhyacophila alticola Kimmins, 1953: 544–545

Adult: Adults large (expanse about 40 mm), medium fuscous, sparsely clothed with golden hairs. Head with a dark fuscous patch between the ocelli. Antennae annulated with dull yellow at the joints. Anterior and median legs fuscous, the tibiae with a narrow, fulvous ring at the preapical spur, posterior legs fulvous. Wings long, rather narrow, and sub-acute apically. Pubescence sparse. The membrane of anterior wings pale fuscous, rather darker along R and Rs, and an oblique dark patch between Cu, and A, about halfway along the wing, and small spots at the bases of several of the forks. Posterior wing hyaline. In both wings, forks 1 and 2 are long, with very short footstalks, R2 + 3 arises at a steep angle from Rs and the radial area is wide. Abdomen yellowish, darker above, anal appendages yellowish-brown. No ventral processes, but sternites VI and VII have each a strong, transverse carina.

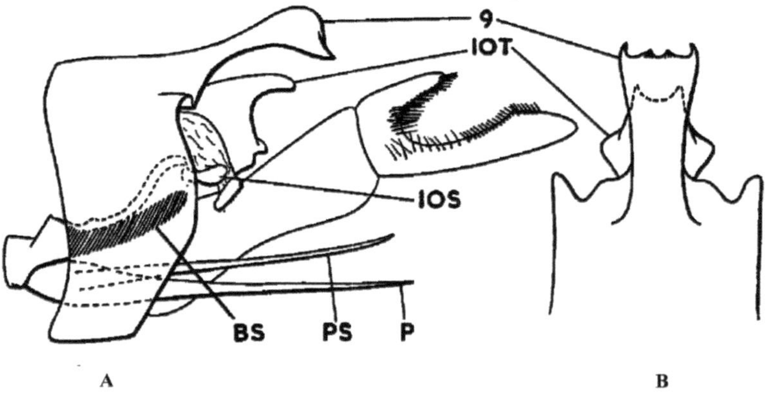

Figure 7.2 *Rhyacophila alticola* (♂): (A) Lateral view of male genitalia; (B) Dorsal view of IXth and Xth tergites (from Kimmins, 1953).

Genitalia (♂): (Figs. 7.2A–B). IXth segment with dorsal margin much produced in a long finger, its apex dilated and with triangular lower angles. Between these triangles are two small, downwardly directed teeth. Xth tergite arises at the base of the dorsal process, about 2/3 of its length. Its apex is excised, and its lateral margins are extended downwards and terminate in a pair of small, elongated, rugose lobes. X^{th} sternite semi-membranous, with a sclerotized apical margin. The basal sleeve of phallus with its upper margin projecting in a long, slender lobe with dilated apex. Phallus and sheaths long and styliform. Inferior appendage long, basal segment extending as far as apex of dorsal process, a little constricted before the middle. Terminal segment about three-quarters as long as basal, excised almost to its base to leave two fingers, the lower stouter and twice as long as upper, inner surfaces spinose.

Genitalia (♀): Not available.

Holotype: TIBET: Gaotsa, 12,300 ft; 21.III.1933, Phari, 14,300 ft; 22.III.1933.

Holotype depository: BMNH, London.

Distribution: Tibet, Nepal, India (Sikkim).

3. ***Rhyacophila anatina*** **Morton**

Rhyacophila anatina Morton, 1900: 6

Adult: Head thorax and abdomen dark fuscous, paler beneath. Apex of wing somewhat elliptical; apical fork no. 2 equal to or longer than no. 1. Costal hairs simple. Membrane fuscous, with golden irrorations; pterostigma dark. Antennae fuscous with pale annulations.

Genitalia (♂): (Figs. 7.3A–B). IXth segment produced in a dorsal process with a stout, bifid apex. On either side of the dorsal process is a broad, oval plate, part of the Xth tergite. Their inner margins meet beneath the dorsal process and are extended downward and then basally. X^{th} sternite membranous. Basal sleeve of phallus large, encircling phallus and sheaths at their bases. Its upper margin is produced in a strong, quadrate hood with a slightly concave apex, reaching almost to the apex of the phallus. The phallus itself is short, slender and spiniform. Sheaths short and stout, apices with a dense tuft of long setae, directed upward. Inferior appendage moderately long, basal segment stout, slightly curved, terminal about half as long and half as wide as basal.

Genitalia (♀): Not available.

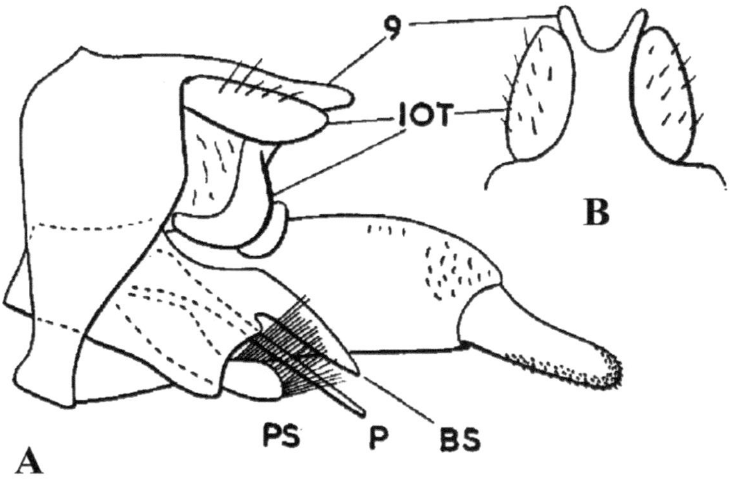

Figure 7.3 *Rhyacophila anatina* (♂): (A) Male genitalia, lateral view; (B) Dorsal view of IXth and Xth tergites (from Kimmins, 1953).

Holotype: INDIA: Assam; Khasi Hills; 6-XII-1899.

Holotype depository: BMNH, London.

Distribution: China, India (Meghalaya, Sikkim).

4. ***Rhyacophila ancestralis* Martynov**

Rhyacophila ancestralis Martynov, 1935: 93–95

Adult: Head and thorax black-brown; antennae and palpi brown; 1st and 2nd joints of the maxillary palpi very short, 3rd joint long, the 5th equal to it, 4th shorter; in the labial palpi, the 2nd joint is equal to the 3rd. Anterior wings brownish, moderately narrow; R thick, brown, ending in an elongated fork, which bounds dark spot (pterostigma); branches of RS also thick, 2nd apical fork beginning much earlier than the 1st, 3rd fork equal in length to the 4th. Posterior wings yellowish; R thick, without apical fork, but the pterostigmal spot also dark; 2nd apical fork beginning much earlier than the 1st, 3rd fork short, M weak. Abdomen brownish above, ochraceous beneath.

Genitalia (♂): (Figs. 7.4A–C). IXth segment yellowish-brown, much broadened in its upper lateral parts; IXth tergite also broad, with an elongated median projection, slightly narrowing backward and truncated at its end; hind portion provided above with some transverse tubercles. Laterally, X^{th} segment is roof-shaped, short, but broad, with two small rounded preanal appendices, each bearing several hairs or bristles; apical portion of Xth dorsal segment is separated by excision from the two small lateral processes. Dorsal part of Xth segment appears above as a process of IXth tergite. Phallus slender, with two elongated thick, but pale titillators, bearing several spinules at their ends above.

Genitalia (♀): Not available.

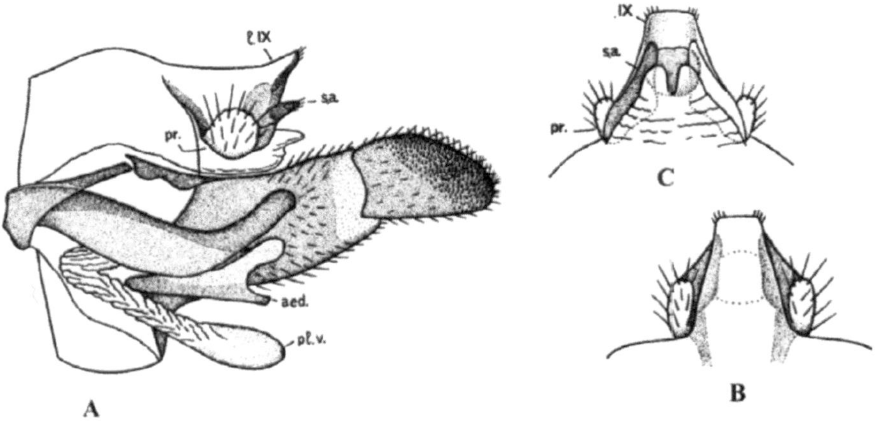

Figure 7.4 *Rhyacophila ancestralis* (♂): (A) Male genitalia, lateral view; (B) Dorsal view; (C) Ventral view (from Schmid, 1970).

Holotype: Dumpep, Khasi Hills, Assam; October, 1929.
Distribution: India (Meghalaya).

5. ***Rhyacophila angden*** Schmid

Rhyacophila angden Schmid, 1970: 183

Adult: Legs and top of the body black-brown, with golden hair. Forewings dark brown, with golden spots, quite distinct and quite numerous. Wingspan: 12.5–17.5 mm. This species is easily distinguished by its mottled forewings and the height of the root of anal sclerites.

Genitalia (♂): (Figs. 7.5A–B). Apical dorsal lobe of the IXth segment wide and bilobed at its end. Anal sclerites with their outer edge raised into a triangular fin; their root is high and angular. Second section of the lower appendages large and triangular.

Genitalia (♀): Not available.

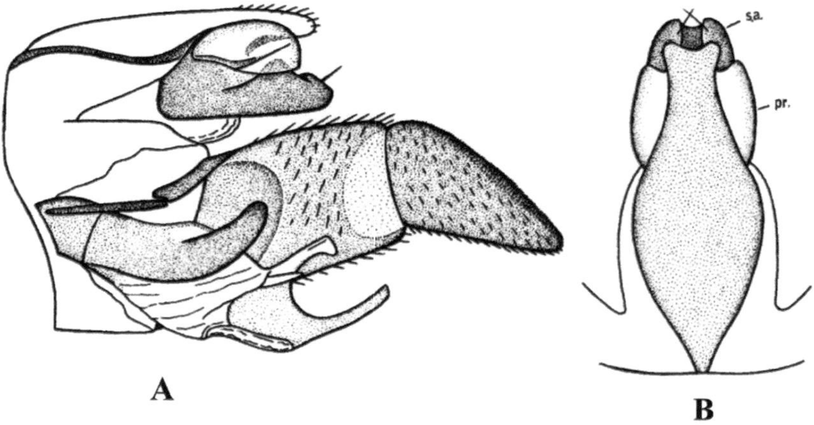

Figure 7.5 *Rhyacophila angden* (♂): (A) Lateral view of male genitalia; (B) Dorsal view (from Schmid, 1970).

Holotype: Sikkim, Chachu; 17-V-1959.

Holotype depository: ROM, Canada.

Distribution: India (Sikkim).

6. *Rhyacophila antiope* Malicky

Rhyacophila antiope Malicky, 1997: 218

Adult: Dark brown, Forewing slightly lighter, speckled on the leading edge.

Genitalia (♂): (Figs 7.6A–C). Xth Segment flat and high; the lateral view has a deep indentation in the middle. The lower appendages are short and high; on the 2nd limb, the ventral lobe is much longer than the dorsal.

Genitalia (♀): Not available.

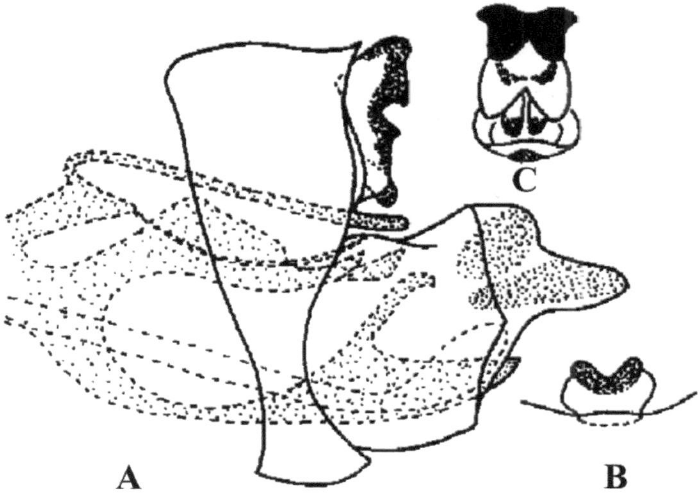

Figure 7.6 *Rhyacophila antiope* (♂): (A) Lateral view of male genitalia; (B) Dorsal view of Xth; (C) Ventral view (from Malicky, 1997).

Holotype: North Andaman, Kalpong, 21-VI-1994.

Holotype depository: ZSI, India.

Additional material examined: INDIA, Andaman Island; Sept. 2020; (Coll. S. H. Parey).

Additional material deposited: Baba Ghulam Shah Badshah University Rajouri, J&K (India).

Distribution: India (North Andaman Island).

7. *Rhyacophila armitageorum* Saini and Kaur

Rhyacophila armitageorum Saini and Kaur, 2012: 311–312

Adult: Head dark brown with golden pubescence. Scape of antenna dark brownish; pedicel and flagellum pale brown alternating light and dark brown color bands on flagellum. Maxillary palp dark brown; Pubescence on all segments of maxillary palp pale brownish except 2nd segment with thick brownish setae. Labial palp light brown. Thorax and abdomen coal brown in color. Foreleg and hindleg light brown but midleg dark brown. Length of antenna 7.3 mm and forewing 9.0 mm and dark brown in color but hindwing light brown. In forewing, fork 1st arises earlier than 2nd. In females, fork 2nd arises slightly earlier than first.

Genitalia (♂): (Figs. 7.7A–C). Laterally, the terminal segment of the inferior appendage rounded in outline. Segment X hanging downward and anal sclerite rounded. Dorsal appendage of phallic apparatus small and pointed apically; ventral lobe curved; and lateral arm of ventral lobe of phallicata curved upwards; paramere pointed apically; ventrally, the ventral lobe of phallicata W-shaped.

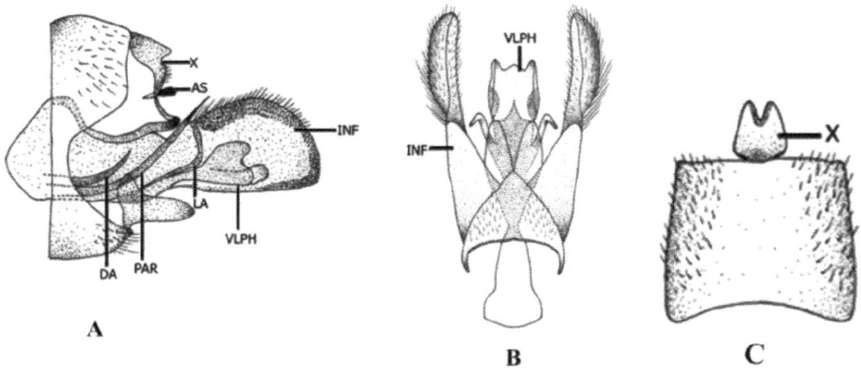

Figure 7.7 *Rhyacophila armitageorum* (♂): (A) Lateral view of male genitalia; (B) Ventral view; (C) Dorsal view.

Genitalia (♀): (Figs. 7.8A–E). Segment VIII with V-shaped indentation and long setae along ventrolateral aspect; dorsal apodeme extending anterad to segment V, basal apodeme extending to segment VI and curved apically. Dorsally and ventrally segment VIII with V-shaped notch apically but long setae present on sternum only. Vaginal apparatus small and simple with rounded apex in lateral view and in ventral view eye-shaped sclerite in posterior process.

Holotype: INDIA: Arunachal Pradesh: Lumla; 2300 m; 15-V-2011. (Coll. Parey & Pandhar).

Holotype depository: MDZPUP, India.

Distribution: India (Arunachal Pradesh).

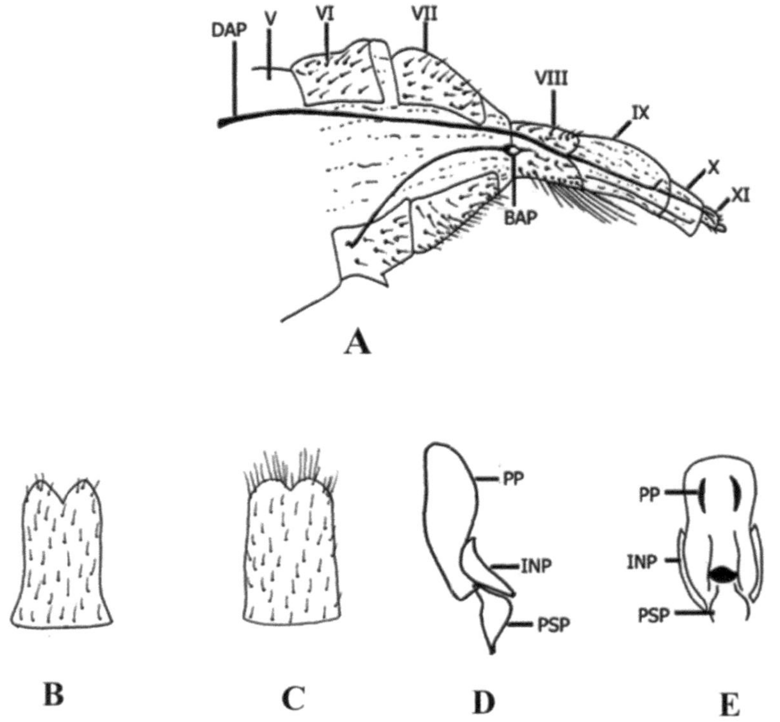

Figure 7.8 *Rhyacophila armitageorum* (♀): (A) Lateral view of the female genitalia; (B) Dorsal view of segment VIII; (C) Ventral view of segment VIII; (D) Lateral view of vaginal apparatus; (E) Ventral view of vaginal apparatus.

8. *Rhyacophila assimilis* Kimmins

Rhyacophila assimilis Kimmins, 1953: 553

Adult: Small-sized insect, body piceous, legs fuscous. Wings with reddish fuscous pubescence, membrane brownish, with white cross-veins and forks.

Genitalia (♂): (Figs. 7.9A–B). Abdomen with a strong, triangular process on VIIth sternite and a shallow transverse ridge on the VIth sternite, giving the appearance of a small process in lateral view. IXth segment with the lateral margins less convex, Xth segment with its upper margin straight, lower apical angle less produced downwards, apex more excised. Phallus with the upstanding lobes on the basal sleeve larger. Sheaths almost straight. Phallus itself without the deep, ventral keel, membranous ventral lobe larger. Inferior appendage with first segment relatively longer, about three times as long as wide, terminal segment with the apical excision less wide, upper branch larger. Spinules present on the upper surfaces of both branches.

Genitalia (♀): Not available.

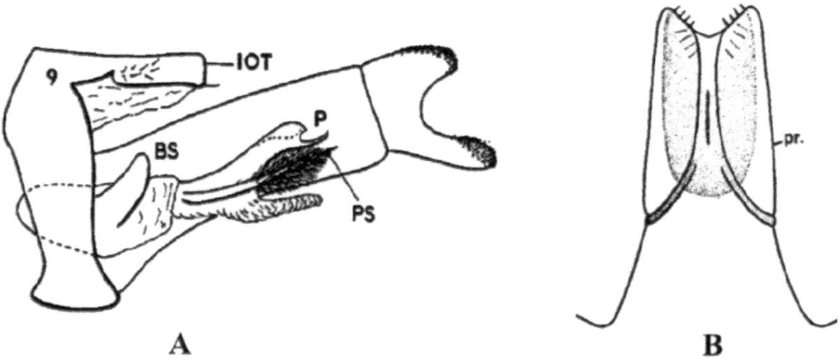

Figure 7.9 *Rhyacophila assimilis* (♂): (A) Male genitalia, lateral view; (B) Dorsal view of IXth and Xth segments (from Kimmins, 1953).

Holotype: N.E. Burma: Kambaiti, 1800–2000 ft., 21-V, 17-VI-1934.

Holotype depository: Stockholm Museum, Sweden.

Distribution: Myanmar, India (Manipur).

9. *Rhyacophila aureomaculata* Schmid

Rhyacophila aureomaculata Schmid, 1970: 184

Adult: Entire body dark brown. Length of antenna 5.5 mm and forewing 6.79 mm. In forewing, fork 2nd arises earlier than 1st.

Genitalia (♂): (Figs. 7.10A–C). In dorsal view apex of anal sclerite rounded, elongated and stalked. Preanal appendage extends up to the stalk of anal sclerite. Dorsal apical lobe of segment IXth segment truncates apically. In lateral view segment IX long and broad. Dorsal apical lobe of segment IX rounded. Preanal appendage triangular shaped. Anal sclerite sub-rectangular. Phallic apparatus with phallicata long and upcurved. Ventral lobe of phallicata shorter than phallicata. Dorsal appendage of phallic apparatus rounded. Terminal segment of inferior appendage three times smaller than the basal segment. Basal segment broad at the base and terminal segment of inferior appendage with minute spinules. In ventral view, basal segment of inferior appendage convexly raised near the base.

Genitalia (♀): Not available.

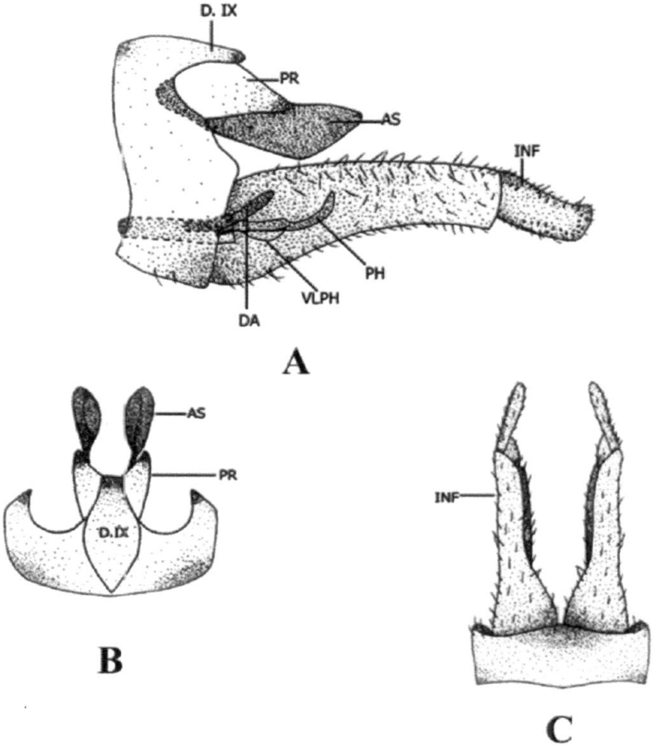

Figure 7.10 *Rhyacophila aureomaculata* (♂): (A) Lateral view of male genitalia; (B) Dorsal view; (C) Ventral view.

Holotype: INDIA: Assam; NEFA, Kameng Frontier Division, Shergaon, 8-v-1961.

Material depository: ROM, Canada.

Additional material examined: INDIA: Arunachal Pradesh: Menchuka, 2300 m, 29-iv-2010. (Coll. Parey & Pandhar).

Distribution: India (Arunachal Pradesh).

10. ***Rhyacophila aureostigma*** **Schmid**

Rhyacophila aureostigma Schmid, 1970: 184

Adult: Upper body dark brown, with light reddish-brown legs. Upper abdomen light brown; orange-yellow underside and genitals. Forewings dark brown, with a shiny golden patch behind the pterostigma and different in size in both sexes. In male, it extends into the end of the radial cell and into the first two apical cells; it overflows only rarely and only slightly in the third apical cell; sometimes, but rarely, it reaches the apex of the wing. In female, it has the same limits as in male, but in addition covers all the pterostigma. Wingspan 9: 11–17 mm.

Genitalia (♂): (Figs. 7.11A–B). Preanal appendages in narrow ovals, less high than the anal sclerites and with their lower edge a little concave. Anal sclerites large, in the form of regular parallelograms, high, massive with their lower apical angle a little prominent. This species is very similar to *aureomaculata* but can be recognized by the golden patch on the smaller forewings and the height of the anal sclerites.

Genitalia (♀): Not available.

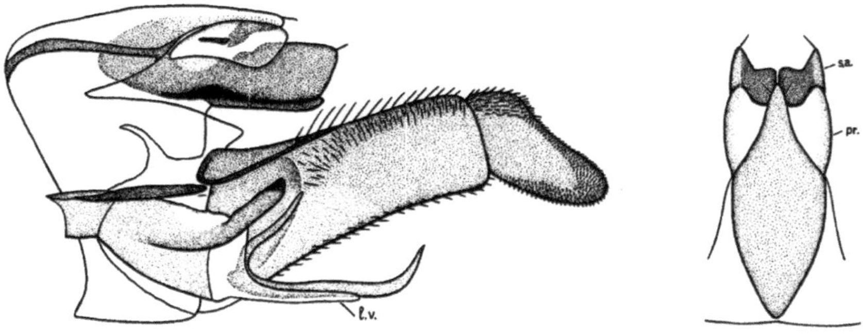

A B

Figure 7.11 *Rhyacophila aureostigma* (♂): (A) Male genitalia, lateral view; (B) Dorsal view (from Schmid, 1970).

Holotype: INDIA, Western Bengal, Dilpa, 21-III-1959.

Material depository: ROM, Canada.

Distribution: Nepal, India (West Bengal).

11. *Rhyacophila bhotia* Schmid

Rhyacophila bhotia Schmid, 1970: 137

Adult: Upper body dark brown, with golden hairiness; pale reddish paws. Forewings dark brown, uniformly, and densely riddled with fine golden macules. Wingspan: 13–19 mm. This species is isolated and could be the most primitive of the group by the regular proportions of the lower appendages (Schmid, 1970).

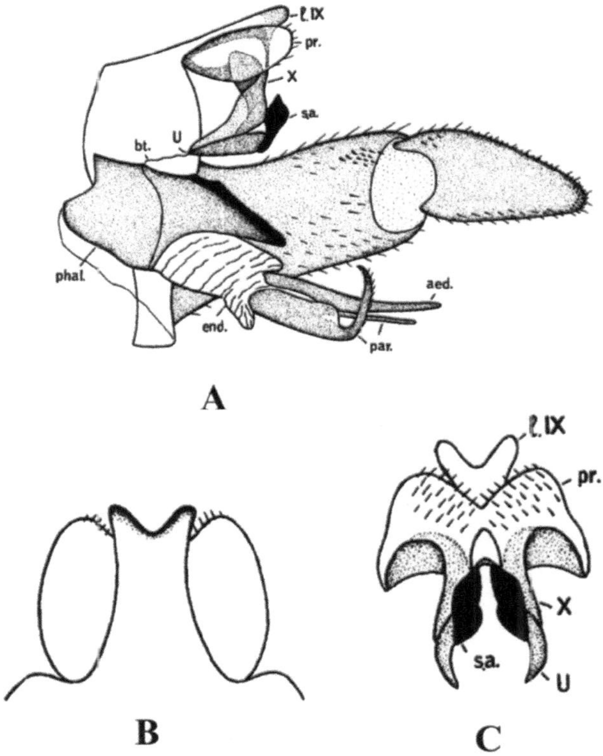

Figure 7.12 *Rhyacophila bhotia* (♂): (A) Lateral view of male genitalia; (B) Dorsal view; (C) Dorsal view of Xth segment (from Schmid, 1970).

Genitalia (♂): (Figs. 7.12A–C). Preanal appendages appearing ovoid when viewed from above. Lower appendages of medium size and slightly thickened; second article quite large and triangular. Dorsal appendage of the phallic apparatus thick and entire. Parameres short, thick and terminated in two branches perpendicular to each other.

Genitalia (♀): Not available.

Holotype: INDIA: Tehri Garhwal, Dobalgaon; 14-IV-1958.

Material depository: ROM, Canada.

Distribution: India (Uttarakhand).

12. *Rhyacophila bhuchanadhara* Schmid

Rhyacophila bhuchanadhara Schmid, 1970: 210

Adult: Head, antenna, maxillary palp, labial palp and wings pale brown; prothorax and metathorax brown, mesothorax dark brown; forewing with dark brown irrotations; fore and midlegs dark brown, hindleg orangish. Length of antenna 9.21 mm and forewing 11.64 mm. In forewing forks 1st and 2nd arise at the same level.

Genitalia (♂): (Figs. 7.13A–C). In dorsal view dorsal apical lobe of segment IX indented. In lateral view dorsal apical lobe of segment IX hood shaped and cover segment X and anal sclerite. Segment X with a claw-like appendage at apical margin. Anal sclerite rounded in shape. Dorsal appendage of phallic apparatus curved and rounded apically. Spine-like paramere present and it is very much shorter than phallicata. Phallicata broader at middle and near margin it is with membranous bulbous structure. Basal segment of inferior appendage indented. In ventral view basal segment of inferior appendage with U shape indentation and terminal segment rounded.

Genitalia (♀): Not available.

Holotype: INDIA: Assam, NEFA, Kameng Frontier Division, Lifakpo: 15-III-1961.

Material depository: ROM, Canada.

Additional material examined: INDIA: Arunachal Pradesh: Lumla; 2300 m; 15-v-2011. (Coll. Parey & Pandhar).

Distribution: Bhutan, India (Arunachal Pradesh).

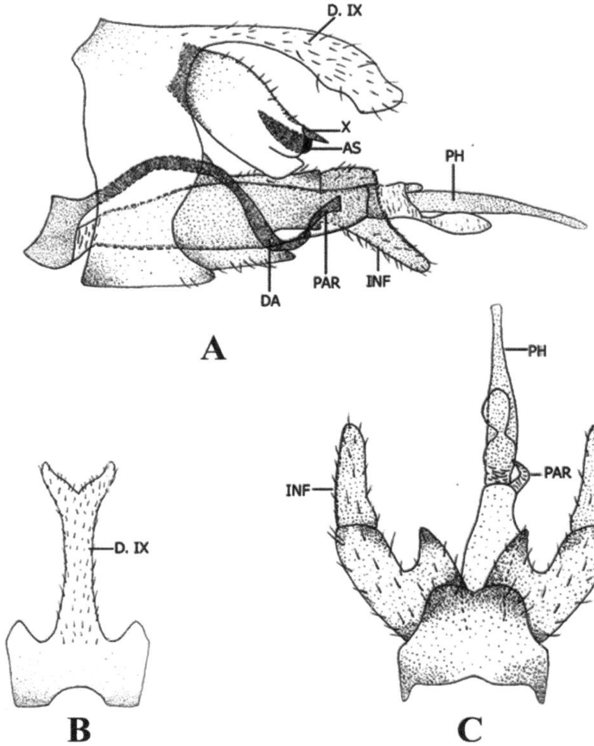

Figure 7.13 *Rhyacophila bhuchanadhara* (♂): (A) Lateral view of male genitalia; (B) Dorsal view; (C) Ventral view.

13. *Rhyacophila bicolor* Kimmins

Rhyacophila bicolor, Kimmins, 1953: 532–534

Adult: Head, thorax and legs fuscous, abdomen dark orange. Wings with fuscous pubescence, in forewing also with white pubescence in a rounded area covering part of apical cellules 4 and 5, and in a narrow, oblique band in the sub-costal and radial areas, from just beyond the fork of R_s towards the base of the pterostigma. Membrane of forewing pale brownish, with darker venation. There is a white thyridial spot, the apex of Cu_2 is white as is also the membrane under the area of white pubescence. Hind wing membrane pale brownish, cross-vein R-M, fork of M and cross-vein M-Cu, part of the stem of M and of Cu_2 and an area in the apical cellules 4 and 5 white.

Genitalia (♂): (Figs. 7.14A–C). IXth segment with its dorsal surface elevated in a rounded, longitudinal hump, dorsal margin triangularly produced and transparent. Lateral margins each with an acute notch. Xth tergite composed of a pair of broad, convex plates, cordate in side view; between these plates are two small, spatulate processes. Xth sternite membranous. Phallus sheaths simple, upcurved and blunt from the side, acute and incurved from beneath. Phallus short and complex; there is an upper, transverse plate, notched at its center to form two lobes, between which is a minute, upcurved spine. From beneath the plate arises a stronger, curved, spine-like duct and below this another plate, terminating in two acute teeth. Inferior appendages elongate, extending a little beyond Xth tergites. Its basal segment is about twice as long as terminal, which is slightly narrower and with an obliquely truncate apical margin.

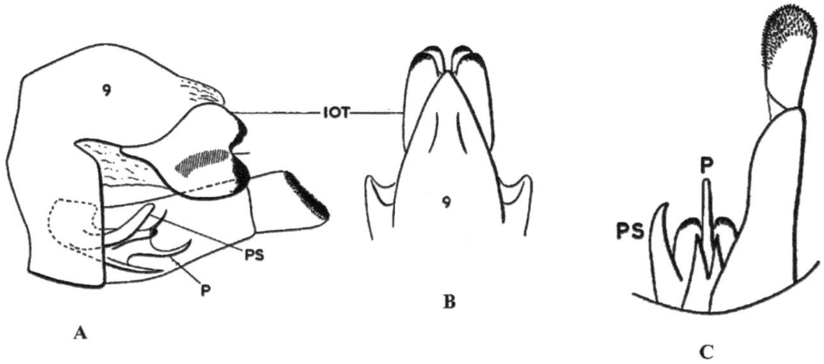

Figure 7.14 *Rhyacophila bicolor* (♂): (A) Male genitalia, lateral view; (B) Dorsal view of IXth and Xth tergites; (C) Ventral view of inferior appendage, phallus and sheaths (from Kimmins, 1953).

Genitalia (♀): (Figs. 7.15A–B). Abdomen with a strong process to the VIth sternite, none to the VIIth segment. VIIIth segment divided into tergite and sternite only in apical half. Apical margin excised dorsally to about half the depth of the segment, and ventrally with only a small emargination.

Holotype: INDIA: N. E. Burma, Kambaiti, 1800–2000 m, 30-1V, 11, 12, 28-V, 11, 17-V1-1934.

Material depository: Stockholm Museum, Sweden.

Distribution: Myanmar, Vietnam, India (Manipur).

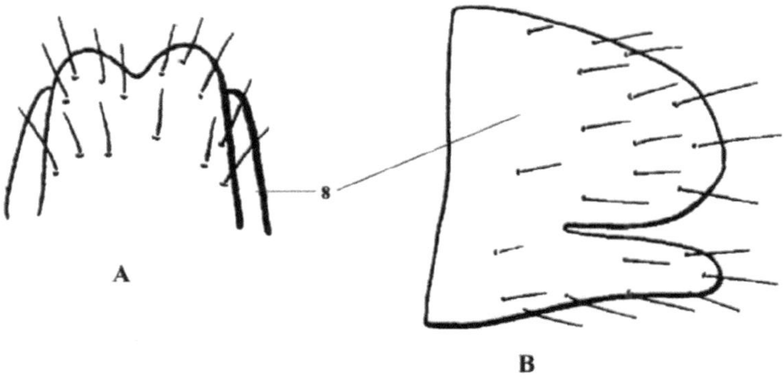

Figure 7.15 *Rhyacophila bicolor* (♀): (A) Ventral view of VIIIth segment; (B) Lateral view of VIIIth segment (from Kimmins 1953).

14. *Rhyacophila bidens* Kimmins

Rhyacophila bidens, Kimmins, 1953: 521

Syn. *Rhyacophila szaboi* Kiss, 2013

Adult: Head and thorax dark brown; antenna, maxillary palp, labial palp, abdomen, wings and legs pale brown but tibia of hindleg with brown distinct marking. Length of antenna 6.79 mm and forewing 7.3 mm. In forewing fork 2nd arise slightly earlier than 1st. In **females**, entire body brown; antenna with alternating dark and light brown bands; legs pale brown, tibia and tarsus of fore and midleg entirely dark brown but only apices of tibia and tarsus of hindleg dark brown. Length of antenna 6.9 mm and forewing 9.7 mm.

Genitalia (♂): (Figs. 7.16A–C). In dorsal view segment X bilobed. IXth segment narrow, indented at the attachment of inferior appendage. Inferior appendage two-segmented, basal segment of wide at its base. Terminal segment large dilating to an oblique excised apex, lower lobe extending beyond superior lobe. Both superior as well as inferior lobes of second segment with minute spinules. Tergite X short, with the apical angles forming two divergent teeth and enclosing a wide excision. Anal sclerite small placed at the base of segment X. Phallicata long and bifid forming two arms. Apex of paramere clavate and setiferous. In ventral view base of the terminal segment of inferior appendage broader than its apex.

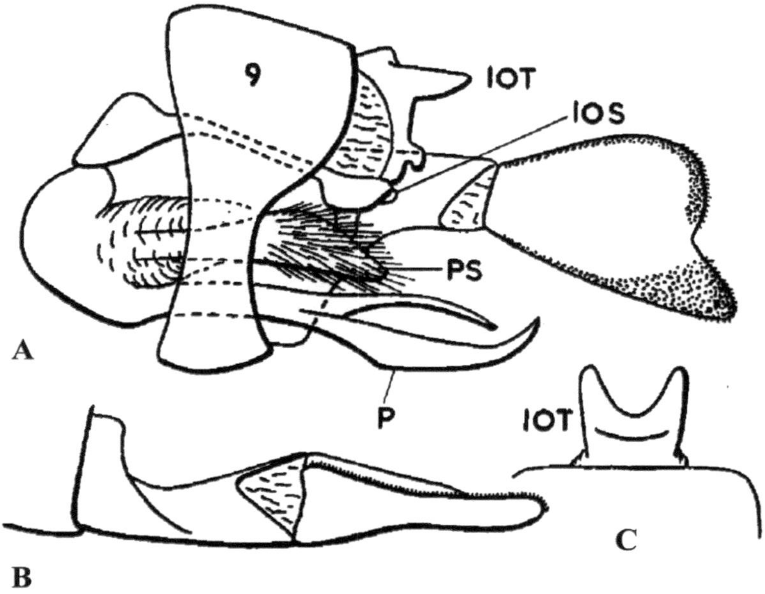

Figure 7.16 *Rhyacophila bidens* (♂): (A) Male genitalia, lateral view; (B) Ventral view of inferior appendage; (C) Dorsal view of Xth tergite (from Kimmins, 1953).

Genitalia (♀): (Figs. 7.17A–E). Apodemes extending anteriorly until segment VII. Intersegmental membranes between VI–VII, VII-VIII and VIII–IX, capable of protraction and retraction. Laterally, dorsal margin of segment VIII curved at middle and rounded posteriorly. Ventrallt, intermediate process of vaginal apparatus divided into two lobes which are free at tip and afterwards fused along the central line.

Holotype: N.E. Burma: Kambaiti, 7000 ft., 2000 m; 28-V, 12-22-VI-1934.

Holotype depository: Stockholm Museum, Sweden.

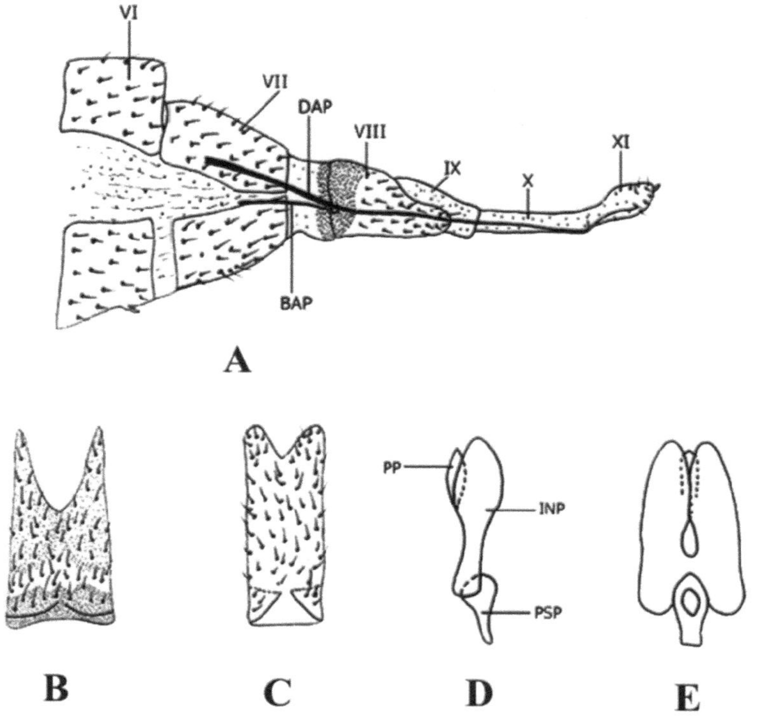

Figure 7.17 *Rhyacophila bidens* (♀): (A) Lateral view of female genitalia; (B) Dorsal view of segment VIII; (C) Ventral view of segment VIII; (D) Lateral view of vaginal apparatus; (E) Ventral view of vaginal apparatus.

Additional material examined: INDIA: Jammu & Kashmir: Varinag, 2300 m, 12.viii.2009; Aru, 2400 m, 13.viii.2010; Uttarakhand: Hanuman Chatti, 3200 m, 25.ix.2008; Shyna Chatti, 2000 m, 27.ix.2008; Tala, 1800 m, 12.vi.2009; Pothibasa, 2200 m, 15.vi.2010; Sagar, 1400 m, 14.vi.2011; Badrinath, 3400 m, 22.vi.2011; Gobinddham, 3100 m, 18.vi.2012; Arunachal Pradesh: Tato, 1700 m, 28.iv.2010; Pene, 1200 m, 30.iv.2010; Bomdila, 2400 m, 03.x.2010; Lumla, 2300 m, 15.v.2011. (Coll. Parey & Pandhar).

Distribution: Bhutan, Myanmar, Nepal, Vietnam, Pakistan/ India (Arunachal Pradesh, Jammu & Kashmir, Manipur, Sikkim, Uttarakhand).

15. *Rhyacophila chakungpa* Schmid

Rhyacophila chakungpa, Schmid, 1970: 161–162

Adult: Dorsal side of the body black-brown, with concolorous hairiness; dark reddish-brown fore and middle legs. Posterior hips, femurs, corresponding tibias, and abdomen light orange; Darker posterior tarsi. Wings uniformly black. Wingspan: 18–19 mm.

Genitalia (♂): (Figs. 7.18A–B). Back piece with a simple, non-notched roof; on its lower face, it is armed with five tiny chitinous teeth. Sub-quadrangular anal scleritis and terminated in three very little protruding lobes. Sagittal appendix in very wide triangle. Second section of the lower appendages triangular, very short and twice as high as long.

Genitalia (♀): Not available.

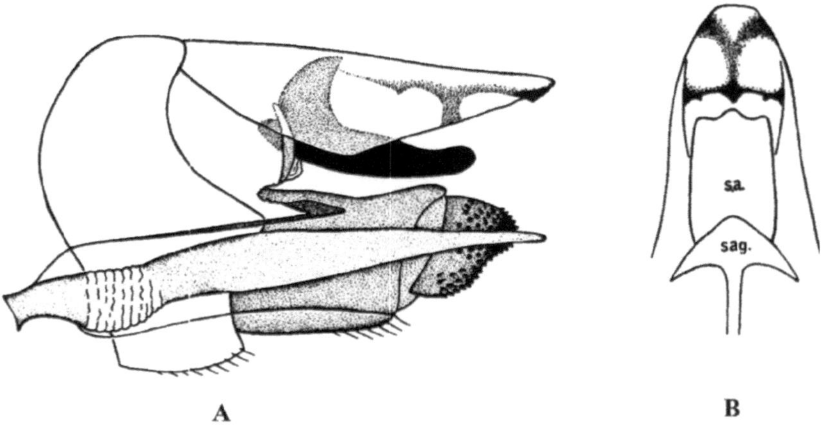

Figure 7.18 *Rhyacophila chakungpa* (♂): (A) Lateral view of male genitalia; (B) Dorsal set seen from below (from Schmid, 1970).

Holotype: INDIA: Sikkim, Rapham, 5250 ft. 2-IV-1959.

Material depository: ROM, Canada.

Distribution: Nepal, India (Sikkim).

16. *Rhyacophila chamolungpa* Schmid

Rhyacophila chamolungpa, Schmid, 1970: 180

Adult: Upper body dark brown, with concolorcus hairiness; reddish-brown paws. Wings black, with an apical hyaline spot, circular and very clear, smaller on the posterior than on the anterior. Bright orange abdomen. Wingspan: 17 mm.

Genitalia (♂): (Figs. 7.19A–C). IXth segment strongly developed at its apical dorsal part in a complex structure and strongly concave. Xth segment of equally complex structure. Phallus composed of a dorsal branch in the roof, covering a cylindrical ventral branch; very large parameres, in spiniferous cones and erectile at their base. Second section of the lower appendages in a rectangle triangle, with its apical concave edge.

Genitalia (♀): Not available.

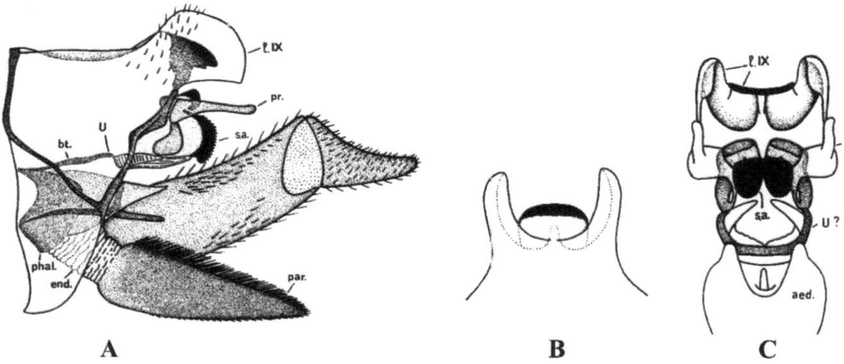

Figure 7.19 *Rhyccophila chamolungpa* (♂): (A) Lateral view of male genitalia; (B) Apical dorsal lobe of IXth segment; (C) Front view (from Schmid, 1970).

Holotype: INDIA: Assam, NEFA, Kameng Frontier Division, Lifakpo, 3100 ft., 15-III-1961.

Material depository: ROM, Canada.

Distribution: India (Arunachal Pradesh).

17. *Rhyacophila chandzo* Schmid

Rhyacophila chandzu, Schmid, 1970: 154

Adult: Head, antenna, maxillary palp, labial palp and wings pale brown; prothorax and metathorax brown, mesothorax dark brown. Wingspan 9: 18–25 mm.

Genitalia (♂): (Figs. 7.20A–B). Vertical part of the Xth segment as long as the horizontal part of the segment; the upper point is in general entirely doubled by a median incision, but its length and its thickness, as well as the depth of the notch, are little variable; robust, odd and simple lower tip. Inferior appendages longer than wide, provided with an apical notch of somewhat variable depth, but always very light and isolating two apical lobes of sub-equal length; the apex of the piece is usually as high as its base.

Genitalia (♀): Not available.

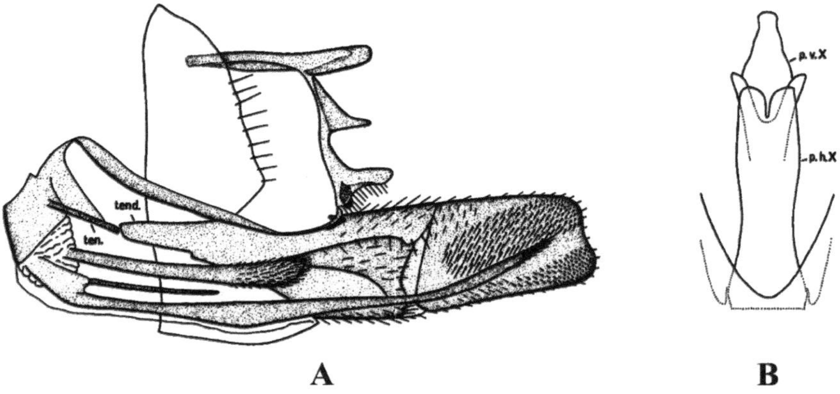

Figure 7.20 *Rhyacophila chandzo* (♂): (A) Male genitalia, lateral view; (B) Dorsal view of Xth segment (from Schmid, 1970).

Holotype: INDIA: Assam, NEFA, Kameng Frontier Division, Jhum La 1-2-VI-1961.

Material depository: ROM, Canada.

Distribution: Nepal, India (Arunachal Pradesh).

18. *Rhyacophila changpa* Schmid

Rhyacophila changpa, Schmid, 1970: 217–218

Adult: Body entirely reddish yellow. Very long and slender antennae, palps and legs. This species is the only one of the genus whose tarsi are shortened and strongly thickened; the fourth article is cordiform; the fifth is devoid of hairiness on its internal surface, which is concave, except at the anterior internal angle which bears a densely hairy prominence

(Schmid, 1970). Long and narrow wings. In the anterior, the costal vein has a fringe of spiky hairs, especially long at the base of the wing. The forewings are light yellow, finely and intensely mottled with brown, with some dark and light areas, quite large in the post-costal area. Wingspan: 21–28 mm.

Genitalia (♂): (Figs. 7.21A–C). Apical dorsal lobe of the IXth segment relatively long and narrow. Xth segment appearing sub-quadrangular and quite angular. Lower (Inferior) appendages high at its end; with a notch at an acute angle, the edges of which are lined with a narrow row of red tubercles and of relatively large size; apical lobe not very stretched.

Genitalia (♀): Not available.

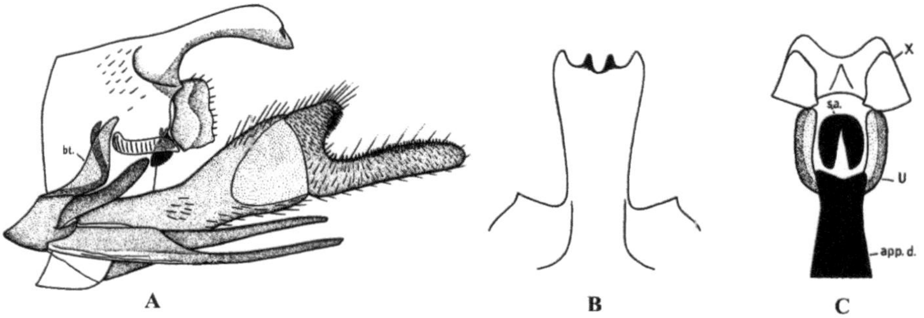

Figure 7.21 *Rhyacophila changpa* (♂): (A) Lateral view of male genitalia; (B) Dorsal apical lobe of IXth segment; (C) Ventral view of Xth segment, tergal set and dorsal appendix of the phallic apparatus (from Schmid, 1970).

Holotype: INDIA: Sikkim, Bakkim, 7000 ft., 12-IV-1959.

Material depository: ROM, Canada.

Distribution: Nepal, India (Sikkim).

19. *Rhyacophila chayulpa chayulpa* Schmid

Rhyacophila chayulpa chayulpa, Schmid, 1970: 190–191

Subspecies: *Rhyacophila chayulpa ringmo,* Schmid, 1970

Subspecies: *Rhyacophila chayulpa tsetangpa,* Schmid, 1970

Adult: Entire body brown. Length of antenna 7.7 mm and forewing 9.7 mm. In forewing fork 2nd arise earlier than 1st.

Genitalia (♂): (Figs. 7.22A–C). In dorsal view anal sclerite longer than preanal appendage and truncate at its apex. In lateral view segment IX broad with dorsal apical lobe prominent. Inferior appendage with its basal segment broad at base and abruptly constricted near the base, terminal segment short deeply excised apically, superior lobe short and rounded. Inferior lobe longer than the superior lobe. Preanal appendage apically trilobed, upper lobe much sclerotized and longer than median and lower lobes; median lobe rounded apically and anal sclerite hidden beneath lower lobe. Phallicata slender shaped and paramere 1/3rd in length as compared to phallicata and with dense apical bunch of setae. In ventral view superior lobe of terminal segment of inferior appendage indented near base of terminal segment of inferior appendage.

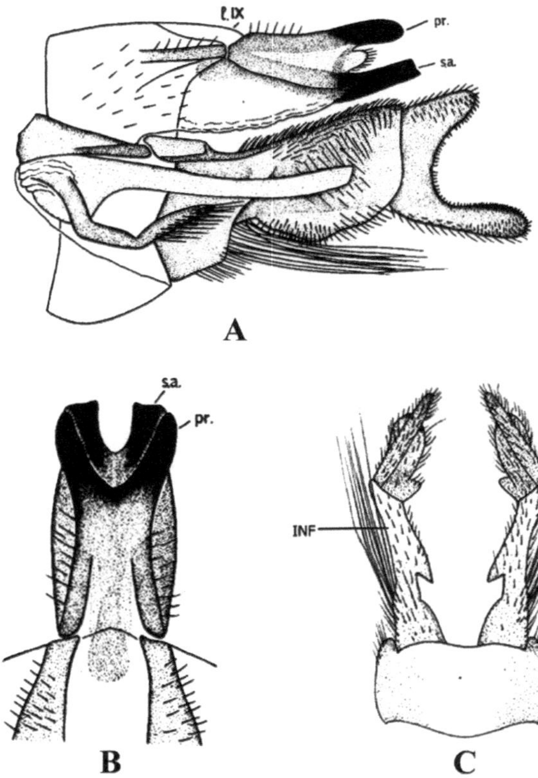

Figure 7.22 *Rhyacophila chayulpa chayulpa* (♂): (A) Lateral view of male genitalia; (B) Dorsal view; (C) Ventral view (from Schmid, 1970).

Genitalia (♀): Not available.

Holotype: INDIA: Sikkim, Lingdok, 7-V-1959.

Holotype depository: ROM, Canada.

Additional material examined: INDIA: Sikkim: Singhik, 1400 m, 14-ix-2009; Uttarakhand: Munsiyari, 2400 m, 17-vi-2008; Munsiyari, 2400 m, 27-vi-2010; Auli, 2900 m, 24-vi-2012. (Coll. Parey & Pandhar).

Distribution: Bhutan, Nepal, India (Sikkim, Uttarakhand).

20. Rhyacophila chematangpa Schmid

Rhyacophila chematangpa, Schmid, 1970: 171–172

Adult: Body reddish-brown, with golden hairiness. Wings uniformly golden yellow. Wingspan: 13 mm.

Genitalia (♂): (Figs. 7.23A–B). Dorsal lobe of the IXth segment narrowed on its apical and bilobed at its end. Clearly bipartite preanal appendages. Dorsal appendages of the phallic apparatus in triangular lobes. Aedeagus (Phallus) long bifid and provided with a deeply indented and chitinous ventral plate on its half apical; robust and setiferous parameres. Inferior appendages sub-quadrangular.

Genitalia (♀): Not available.

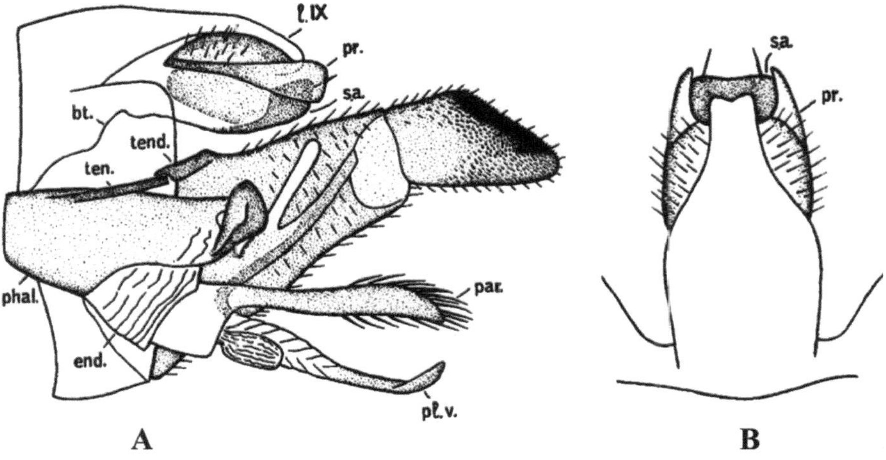

Figure 7.23 *Rhyacophila chematangpa* (♂): (A) Lateral view of male genitalia; (B) Dorsal view (from Schmid, 1970).

Holotype: INDIA: Sikkim, Chematang, 9900 ft., 24-VIII-1959.

Holotype depository: ROM, Canada.

Distribution: India (Sikkim).

21. *Rhyacophila chembo chembo* Schmid

Rhyacophila chembo chembo, Schmid, 1970: 185–186

Subspecies: *Rhyacophila chembo lartsepa,* Schmid, 1970

Adult: Above of the body black-brown, with concolorous hairiness; reddish-brown paws. Forewings black, with two broad patches of light gray hairs at the tips of apical cells 3 and 4 and 6 and 7; their size is variable, and they can sometimes be missing; some specimens also have an area of gray hairs before the arculus. Wingspan: 14–19 mm.

Genitalia (♂): (Figs. 7.24A–B). End of the apical dorsal lobe of the IXth segment clearly visible in lateral view and appearing shortly split when seen from above. Anal sclerites with their external edges raised in a triangular fin and their internal edge also raised in a low keel and of fairly regular height. Phallus very little sclerotized and composed of two sub-equal branches; parameres absent or reduced to the state of short triangles. Inner section of the lower appendages (inferior appendages) with a wide rounded notch. This species is similar to *R. angden* in the shape of anal sclerites but is easily recognized by the two gray spots on the forewings (Schmid, 1970).

Genitalia (♀): Not available.

Figure 7.24 *Rhyacophila chembo* (♂): (A) Lateral view of male genitalia; (B) Dorsal view (from Schmid, 1970).

Holotype: INDIA: Assam, NEFA, Fameng Frontier Division, Jhum La, 24-25-III-1961.

Holotype depository: ROM, Canada.

Distribution: Nepal, Bhutan, India (Arunachal Pradesh).

22. ***Rhyacophila chenmo* Schmid**

Rhyacophila chenmo, Schmid, 1970: 185–186

Adult: Top of the head reddish gray; upper chest black-brown; golden hair on the top of the body; reddish-brown paws. Forewings black-brown, with golden macules sparse or absent over most of the wing surface, but quite numerous along the costal edge; at the base of the wing and all along the post-costal border, they form a cross pattern, of rather complex shape. Wingspan: 21–23 mm.

Genitalia (♂): (Figs. 7.25A–E). Xth segment obtuse; seen from above, it appears to be terminated by quadrangular lobes, separated by an equally angular notch. End of the inner side of the inferior appendages (lower appendages) not very thinned, slightly variable shaped and terminated by two obtuse lobes of sub-equal length.

Genitalia (♀): Not available.

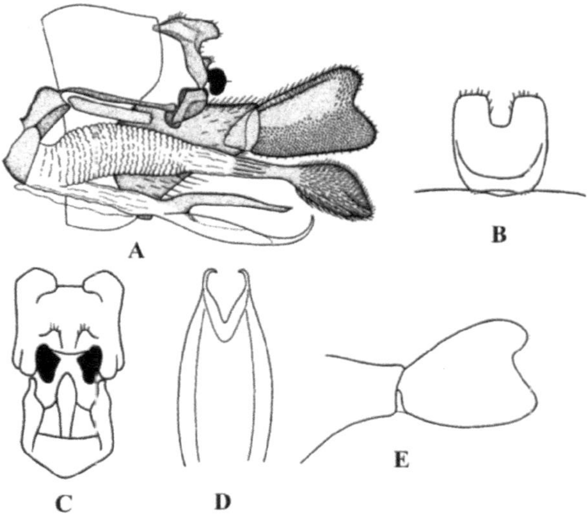

Figure 7.25 *Rhyacophila chenmo* (♂): (A) Lateral view of male genitalia; (B) Dorsal view of Xth segment; (C) Front view of Xth segment; (D) Ventral plate of the phallus; (E) Inner section of Inferior appendage (from Schmid, 1970).

Holotype: INDIA: Manipur, Sirohi Kashong, 7,000 ft., 11-13-VII-1960.

Holotype depository: ROM, Canada.
Distribution: India (Manipur).

23. *Rhyacophila chimdro* Schmid

Rhyacophila chimdro, Schmid, 1970: 172

Adult: Dorsal side of the body black, with concolorous hairiness. Dark reddish-brown legs. Wings uniformly black. Bright orange abdomen. Wingspan: 16–17 mm.

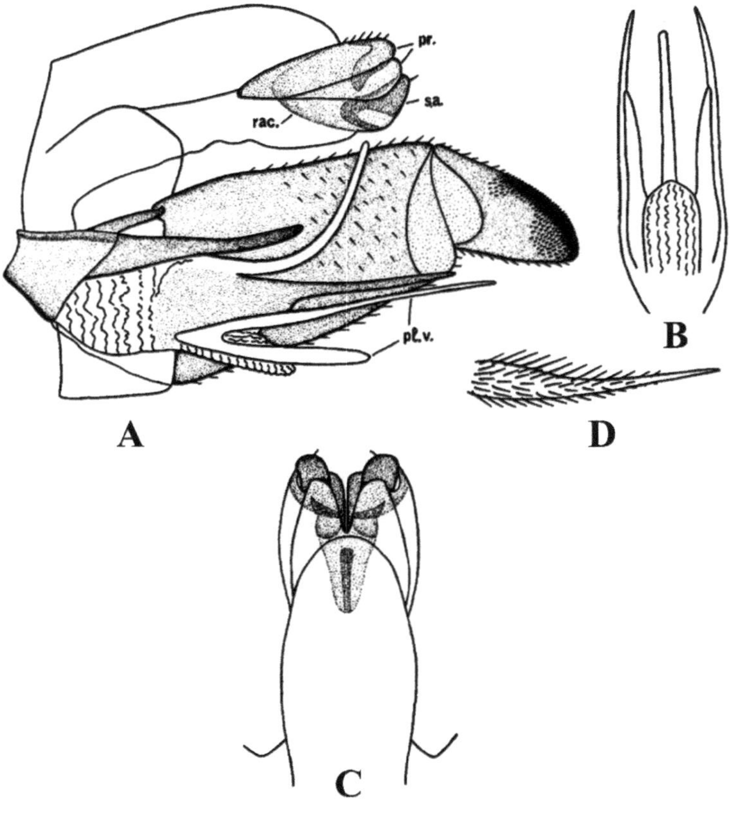

Figure 7.26 *Rhyacophila chimdro* (♂): (A) Lateral view of male genitalia; (B) Ventral view of phallus; (C) Dorsal view; (D) Lateral view of paramere (from Schmid, 1970).

Genitalia (♂): (Figs. 7.26A–D). Apical dorsal lobe of the IXth segment obtuse and strongly domed. Small preanal appendages and anal sclerites. Dorsal appendages of the phallic apparatus in long regular bands. Phallus bifid and accompanied by a ventral plate, also bifid; parameres hairy and very stretched cones. Inner of the Inferior appendages in a triangle with rounded angles. This species is easily recognized by the dorsal appendages of the phallic apparatus in long regular bands and the small size of the preanal appendages and anal sclerites.

Genitalia (♀): Not available.

Holotype: INDIA: Assam, NEFA, Kameng Frontier Division, Bomdi La, 26-30-IV-1961.

Holotype depository: ROM, Canada.

Distribution: India (Arunachal Pradesh).

24. ***Rhyacophila chomoyuma*** **Schmid**

Rhyacophila chomoyuma, Schmid, 1970: 162

Adult: Top of the body reddish-brown. Paws reddish yellow; pinkish orange abdomen. Wings uniformly reddish-brown. Wingspan: 14.5–18 mm.

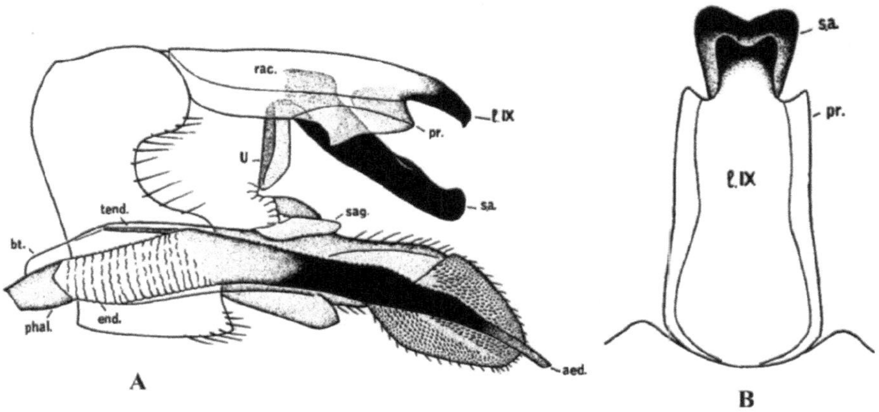

Figure 7.27 *Rhyacophila chomoyuma* (♂): (A) Male genitalia, lateral view; (B) Dorsal view (from Schmid, 1970).

Genitalia (♂): (Figs. 7.27A–B). Apical dorsal lobe of the IXth segment terminated in two circular lobes, slightly curved downwards: preanal appendages clearly visible and forming two apical fins framing the previous ones. Anal scleritis very large and terminated in two circular lobes. Inferior appendages of strong relief, oval distinctly, longer than high. This species is very similar to *R. nakpo*, with which it often cohabits, in the shape of the inferior appendages, but is clearly distinguished by the dorsal part and the anal scleritis.

Genitalia (♀): (Figs. 7.28A–B). VIIIth segment weakly sclerotized. End piece of the copulatory bursa in a very dark cylinder, flattened dorsoventrally, with one of its sides concave.

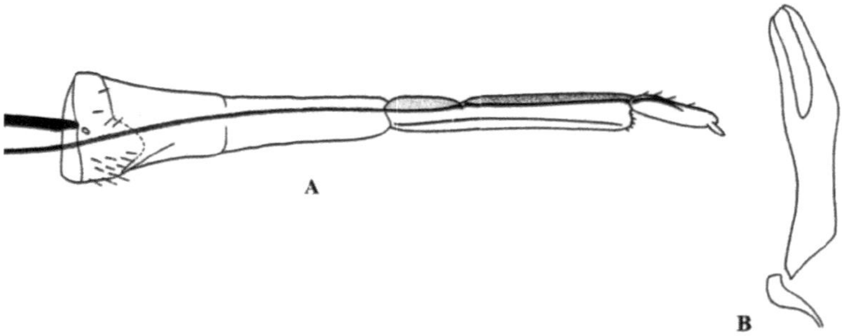

Figure 7.28 *Rhyacophila chomoyuma* (♂): (A) Lateral view of female genitalia; (B) Lateral view of copulatory bursa (from Schmid, 1970).

Holotype: INDIA: Manipur; Hkayam Boum, 3800–8500 ft., 20-23-VI-1960.

Holotype depository: ROM, Canada.

Distribution: India (Manipur).

25. *Rhyacophila choprai* Martynov

Rhyacophila choprai, Martynov, 1935: 97–98

Adult: Head and thorax black; antennae brown; legs brown, clothed with short greyish hairs, spurs long, black. Anterior wings brown, pterostigma dark brown, venation strong, distinct, 2nd apical fork beginning earlier than the 1st, 3rd and 4th forks short enough. In posterior wings 2nd apical fork beginning much earlier, than the 1st, 3rd fork equal in length to its pedicel. Abdomen dark brown.

Genitalia (♂): (Figs. 7.29A–B). Lateral plates of IXth segment very broad, in the form of two broad rectangular plates. IXth tergite separated from these plates by two deep triangular excisions; when seen from above, it is not very broad and long, and by two small lateral constrictions separated from the distal part, composed of a median narrow projection of the IXth tergite and of two lateral elliptic plates; median end-projection of the IXth tergite probably represents morpho logically the Xth tergite united at its base with the IXth; lateral lobes are, perhaps, the preanal appendices. When seen from the side, the lateral lobes are rather narrow, and based from them are seen two triangular figures, margined with brown. Behind and below the lateral plates are situated two more plates, separated by a median excision and forming a sort of furrow. Basal joint of pedes genitals long, broad from side, 2nd small, tongue-shaped from side. When seen from above, basal joint is much narrower, straight; 2nd joint concave internally and bearing here a row of dense short black spinules. (Martynov, 1935).

Genitalia (♀): Not available.

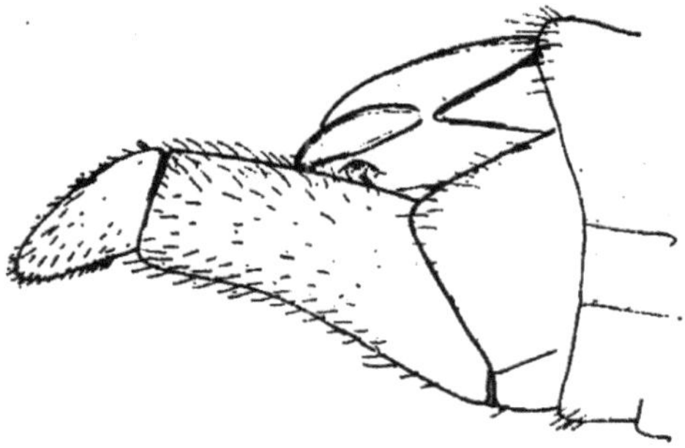

A

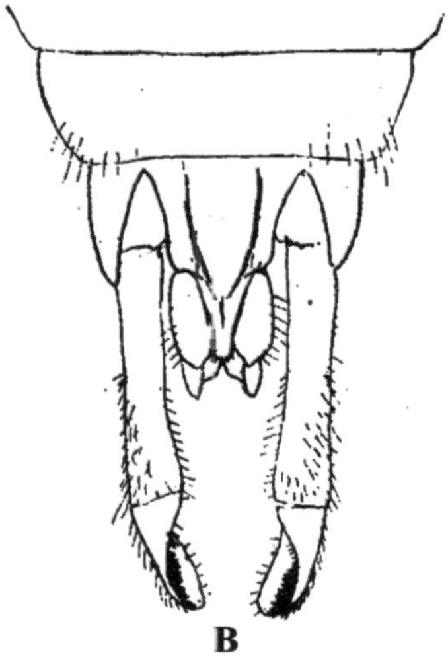

B

Figure 7.29 *Rhyacophila choprai* (♂): (A) Male genitalia, lateral view; (B) Dorsal view (from Martynov, 1935).

Holotype: INDIA: Uttarakhand; Company Khud, below Landour Bazar. Mussoorie, Dehradun district, 18.vi.30.

Distribution: India (Uttarakhand).

26. *Rhyacophila chugalungpa* Schmid

Rhyacophila chugalungpa, Schmid, 1970: 172–173

Adult: Dorsal side of the body black, with concolorous hairiness. Dark reddish-brown legs. Wings uniformly black, except in females, which has a white half-crescent spot on the pterostigma. Bright orange abdomen. Wingspan: 15–17 mm.

Genitalia (♂): (Figs. 7.30A–C). Preanal appendages long, distinctly bilobed ovals. Obtuse and slightly prominent anal scleritis. Dorsal appendages of the phallic apparatus very long and twisted. Phallus composed of a translucent dorsal branch, long oval and concave upwards and a ventral branch in a slender cylinder; ventral plate composed of two strong crossed spines, accompanied by a membranous lower lobe; parameres in long regular bands. This species is obviously related to *R. chimdro* and *R. kyungpa,* but quite distinct by the phallic apparatus (Schmid, 1970).

Genitalia (♀): Not available.

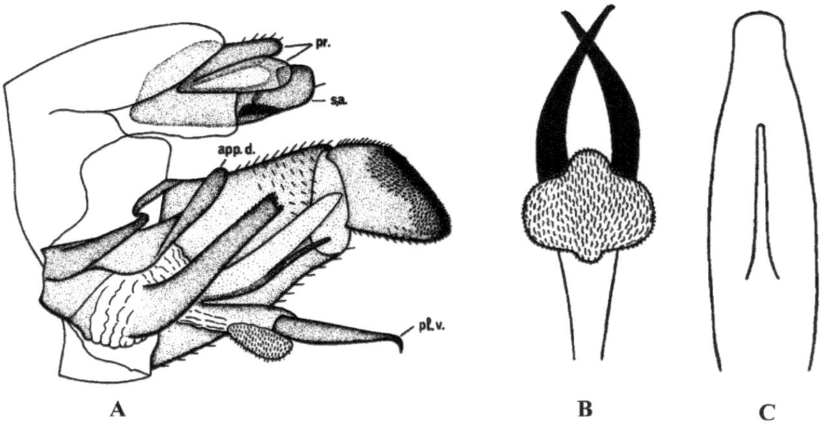

Figure 7.30 *Rhyacophila chugalungpa* (♂): (A) Male genitalia, lateral view; (B) Ventral plate, dorsal view; (C) Dorsal view of phallus (from Schmid, 1970).

Holotype: INDIA: Sikkim; Manu, 5000 ft., 10-V-1959.

Holotype depository: ROM, Canada.

Distribution: India (Sikkim).

27. *Rhyacophila chulukpa* Schmid

Rhyacophila chulukpa, Schmid, 1970: 192–193

Adult: Dorsal side of the body black, brown, with hairs. Dark reddish-brown legs. Bright orange abdomen. Wingspan: 21–24 mm.

Genitalia (♂): (Figs. 7.31A–B). Preanal appendages about twice as long as high and terminated in two rounded and unequal lobes. Phallus strong, sharply thinned before its end which appears as a short slender cylinder; the organ shows two subapical heels, an upper at a right angle and a lower much more obtuse and both covered with dense, fine and short hairiness; parameres long, slender spines. Inferior appendages terminated in two lobes of very unequal size. This species is very close to *R. tarkiya* and is hardly distinguished from it except by the shape of the phallus (Schmid, 1970).

Genitalia (♀): Not available.

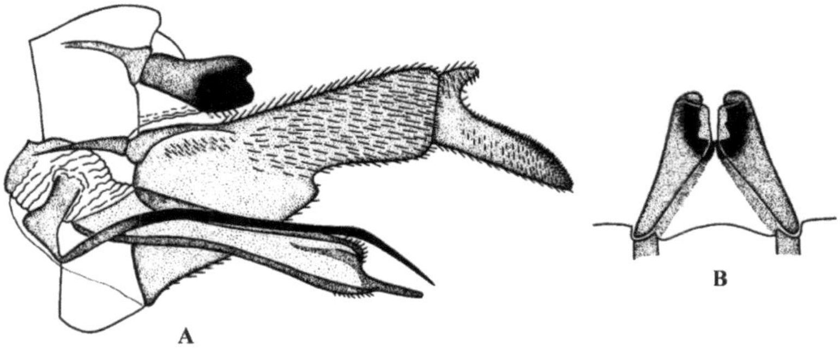

Figure 7.31 *Rhyacophila chulukpa* (♂): (A) Male genitalia, lateral view; (B) Dorsal view of Xth segment (from Schmid, 1970).

Holotype: INDIA: West Bengal; Dilpa, 6200–800 ft., 20-III-1959.

Holotype depository: ROM, Canada.

Distribution: India (West Bengal).

28. ***Rhyacophila chumikpa*** **Schmid**

Rhyacophila chumikpa, Schmid, 1970: 186

Adult: Upper body dark reddish-brown, with concolorous hairiness. Reddish-brown paws. Forewings black, with a circular apical spot, formed by white hairs. Light orange abdomen. Wingspan: 15–19 mm. This species is most often caught with a net and rarely by light (Schmid, 1970).

Genitalia (♂): (Figs. 7.32A–B). Apical dorsal lobe of the IXth segment wide and slightly incised at its end. Long oval preanal appendages, stretched apically and closely joined to the previous one. Anal sclerites long, slender, slightly curved upwards, obliquely truncated at their extremity and a little longer than the dorsal unit. Inferior appendages with the upper apical angle well marked and followed by a clear depression of the apical border.

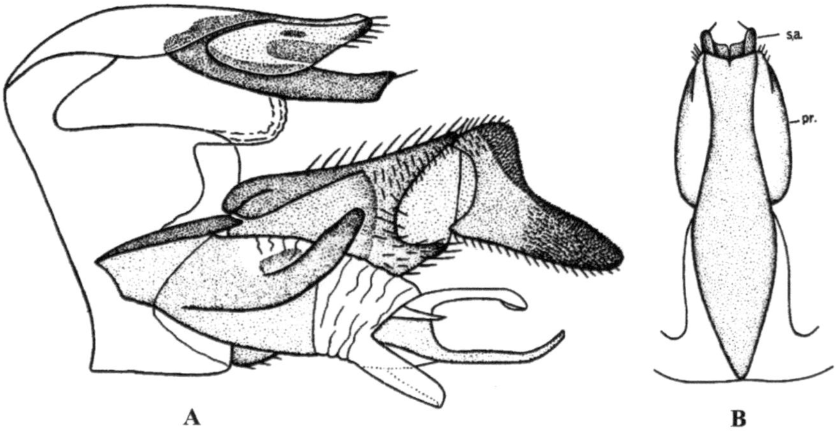

Figure 7.32 *Rhyacophila chumikpa* (♂): (A) Male genitalia, lateral view; (B) Dorsal view (from Schmid, 1970).

Genitalia (♀): (Fig. 7.33). VIIIth segment terminated in three lateral and ventral lobes, rounded at their ends. Terminal part of the copulatory bursa slender, a little dilated at the level of the basal third and the apical third.

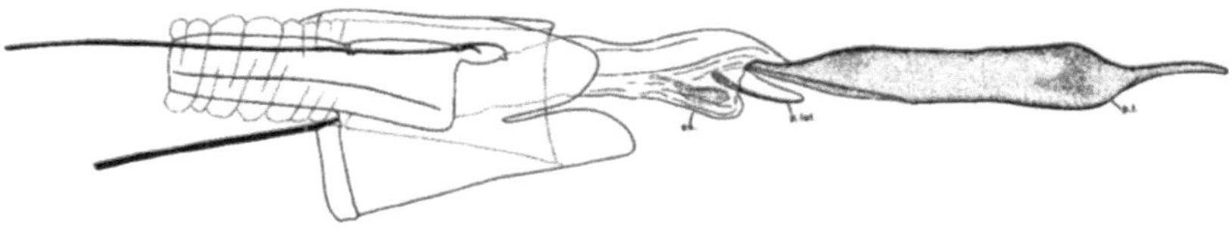

Figure 7.33 *Rhyacophila chumikpa* (♀): Female genital framework (Copulatory bursa) (from Schmid, 1970).

Holotype: INDIA: Assam; NEFA, Kameng Frontier Division, Nyukmadong, 5500–9000 ft., 19-20-IV-1961.

Holotype depository: ROM, Canada.

Distribution: India (Assam, Arunachal Pradesh).

29. *Rhyacophila chungse* Schmid

Rhyacophila chungse, Schmid, 1970: 152

Adult: Entire body brown except antenna, maxillary and labial palpi which are pale brown. Length of antenna 6.79 mm and forewing 8.24 mm. In forewing fork 2nd arise earlier than 1st.

Genitalia (♂): (Figs. 7.34A–D). In dorsal view ends of segment X rounded and in lateral view tergum of segment IX longer than sternum. Segment X slightly indented near base. Anal sclerite rounded apically. Phallicata smaller than ventral lobe of phallicata. Ventral lobe of phallicata broader at base and truncate apically. Paramere bulbous and setifere. Basal segment of inferior appendage broader than terminal one. At base $2^{1/2}$ times broader than apex of basal segment of inferior appendage. Terminal segment of inferior appendage with superior lobe rounded and inferior lobe finger like. Superior as well as inferior lobes with minute spinules. In ventral view basal segment of inferior appendage with deep concavity near apex of basal segment of inferior appendage.

Genitalia (♀): Not available.

Holotype: INDIA: Manipur; Chingsao, 4500–6000 ft., 13-VI-1960.

Holotype depository: ROM, Canada.

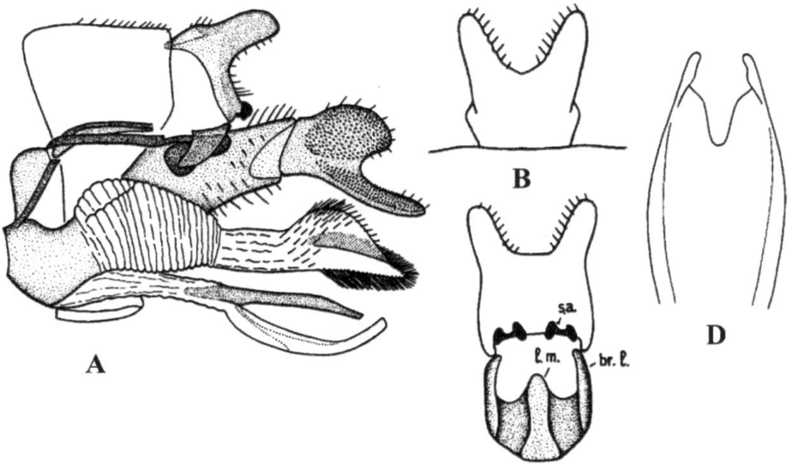

Figure 7.34 *Rhyacophila chungse* (♂): (A) Male genitalia, lateral view; (B) Dorsal view of Xth segment; (C) Front view of Xth segment; (D) Ventral plate of phallus (from Schmid, 1970).

Additional material examined: INDIA: Nagaland: Dzulaki, 1500 m, 06-V-2010. (Coll. Parey & Pandhar).

Distribution: India (Arunachal Pradesh, Manipur, Nagaland).

30. ***Rhyacophila churongpa*** **Schmid**

Rhyacophila churongpa, Schmid, 1970: 176

Adult: Upper body black with concolorous hairiness. Reddish-brown paws. Wings uniformly black. Bright orange abdomen. Wingspan: 14 mm.

Genitalia (♂): (Figs. 7.35A–B). Preanal appendages widely indented, prominent and entirely contiguous to the inferior surface of the apical lobe of the IXth segment. Anal sclerites small and almost completely hidden behind the preanal appendages. Dorsal appendages of the phallic apparatus in very broad and thick crescents, densely furnished with short spines. Phallus small, composed of a short cylinder flanked by two fairly large lateral lobes and armed with asperities; ventral plate not much longer than the phallus and forming two widely spaced points. Inferior appendages tall over their entire length; internally ending in two oval lobes, divergent and well individualized.

Genitalia (♀): Not available.

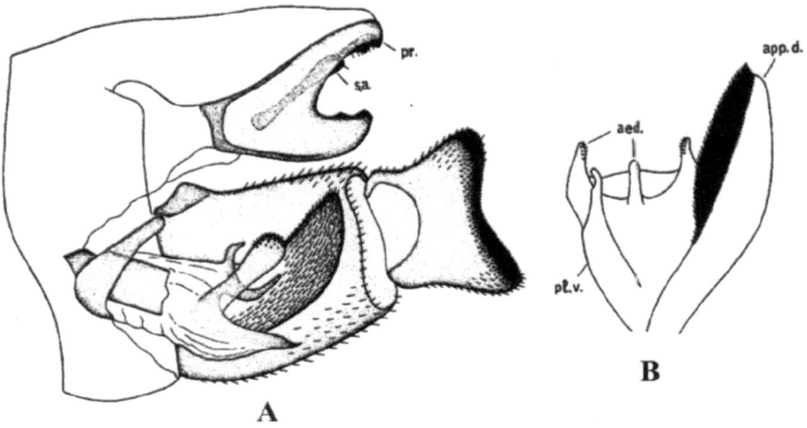

Figure 7.35 *Rhyacophila churongpa* (♂): (A) Lateral view of male genitalia (left Inferior appendage removed); (B) Ventral view of phallic apparatus (from Schmid, 1970).

Holotype: INDIA: Uttarakhand; Almora, Saran, 7700 ft., 17-IX-1958.

Holotype depository: ROM, Canada.

Distribution: India (Uttarakhand).

31. *Rhyacophila curvata* Morton

Rhyacophila curvata, Morton, 1900: 5

Syn. *R. mortoni* Kimmins, 1953

Syn. *R. sumatranu* Ulmer, 1930

Syn. *R. suthepensis* Malicky, 1987

Adult: Entire body dark brown. Length of antenna 8.24 mm and forewing 9.7 mm. In forewing forks 1st and 2nd arise at same level.

Genitalia (♂): (Figs. 7.36A–C). In dorsal view, apices of preanal appendage fused and at base preanal appendage rounded. In lateral view tergum of segment IX broader than sternum.

Preanal appendage broader at apex. Phallicata indented apically and paramere with two teeth one at base and other near the apex. Terminal segment of inferior appendage with superior lobe rounded and inferior lobe finger like. Basal segment of inferior appendage as long as broad. In ventral view terminal segment of inferior appendage with rounded superior lobe and elongated inferior lobe.

Genitalia (♀): Not available.

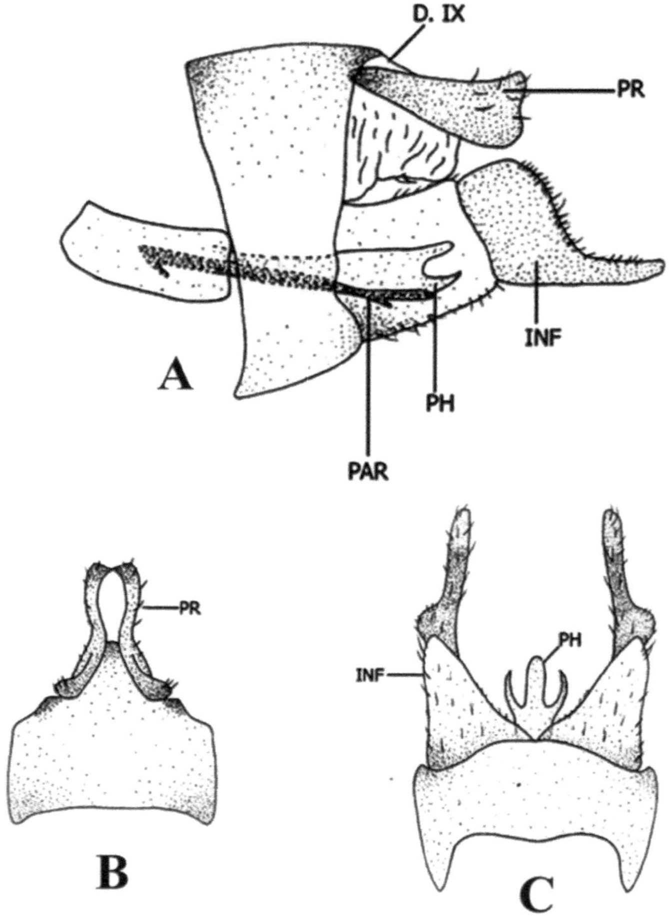

Figure 7.36 *Rhyacophila curvata* (♂): (A) Male genitalia, lateral view; (B) Dorsal view; (C) Ventral view.

Holotype: INDIA: Meghalaya; Khasia Hills.

Holotype depository: BMNH, London.

Additional material examined: INDIA: Sikkim: Phadhamchin, 2200 m, 16-V-2010; Maltin, 1700 m, 20-V-2011; Dentam, 1500 m, 14-V-2011; Meghalaya: Padamamphalang, 1500 m, 26-V-2011. (Coll. Parey & Pandhar).

Distribution: Thailand, China, Indonesia, Malaysia, Myanmar, India (Meghalaya, Manipur, Sikkim).

32. *Rhyacophila dafla* Schmid

Rhyacophila dafla, Schmid, 1970: 189

Adult: Top of the body reddish-brown. Light reddish yellow legs. Plain black-brown wings. Wingspan: 15–19 mm.

Genitalia (♂): (Figs. 7.37A–B). Preanal appendages composed of an upper part in horizontal oval and a lower part, vertical and hyaline. Anal sclerites small, but with a strong root. Phallus in vertical plate, with rounded contours and with a circular apical notch; long and sinuous parameres. Inferior appendages with a large indentation isolating two unequal lobes.

Genitalia (♀): Not available.

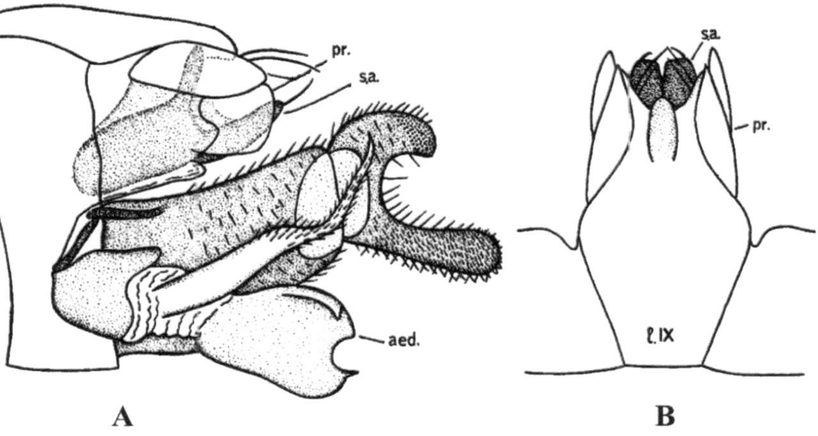

Figure 7.37 *Rhyacophila dafla* (♂): (A) Lateral view of male genitalia; (B) Dorsal view (from Schmid, 1970).

Holotype: INDIA: Assam, NEFA, Kameng Frontier Division, Dirang Dzong, 5300–8800 ft., 9-IV-1961.

Holotype depository: ROM, Canada.

Distribution: India (Arunachal Pradesh).

33. ***Rhyacophila dakshi*** **Schmid**

Rhyacophila dakshi, Schmid, 1970: 186–187

Adult: Head and thorax entirely black, as well as the appendages. Abdomen light orange-pink. Hind legs are somewhat paler than the forelegs. Wings entirely black; no gray spot on the apex of the forewings. This species is very similar to *R. chumikpa* (Schmid, 1970).

Genitalia (♂): (Figs. 7.38A–B). Inferior appendages fairly regularly triangular, with the upper apical angle not protruding and the apical border barely depressed. Preanal appendages generally not stretched at their extremity, nor fused to the apical dorsal lobe of the IXth segment. The latter is of somewhat variable length, anal scleritis also of inconsistent height.

Genitalia (♀): Not available.

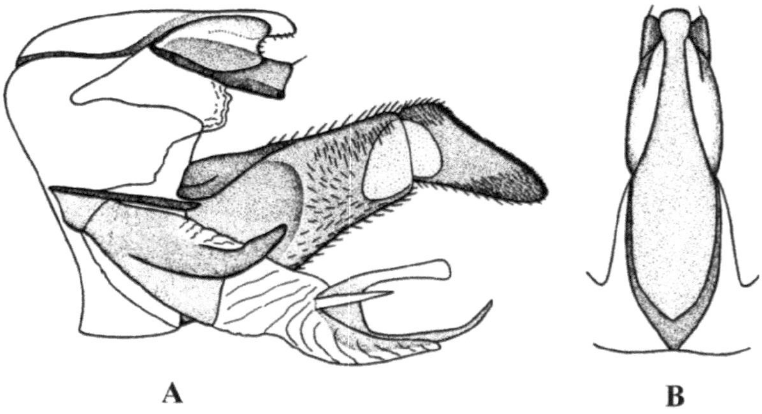

Figure 7.38 *Rhyacophila dakshi* (♂): (A) Male genitalia, lateral view; (B) Dorsal view (from Schmid, 1970).

Holotype: INDIA: West Bengal; Shepi, 4500–8000 ft., 23-III-1959.

Holotype depository: ROM, Canada.

Distribution: Nepal, India (West Bengal).

34. *Rhyacophila dgaldanpa* Schmid

Rhyacophila dgaldanpa, Schmid, 1970: 220

Adult: Entire body brown except maxillary and labial palpi which are pale brown. Length of antenna 7.76 mm and forewing 9.7 mm. In forewing fork 2nd arise earlier than first.

Genitalia (♂): (Figs. 7.39A–C). In dorsal view dorsal apical lobe of segment IX membranous from middle to apex and at middle dorsal apical lobe of segment IX with deep notch and rounded setiferous lobes are present. In lateral view, tergum of segment IX longer than sternum. Dorsal apical lobe of segment IX roof like and cover all beneath lying appendages except anal sclerite which is longer than it. Segment X beneath dorsal apical lobe of segment IX. Dorsal appendage of phallic apparatus longer than ventral lobe of phallicata.

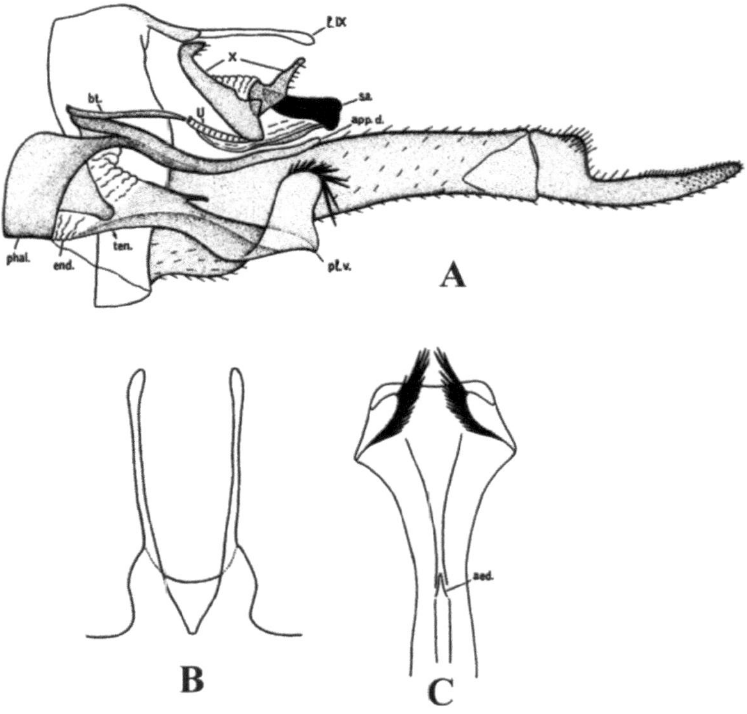

Figure 7.39 *Rhyacophila dgaldanpa* (♂): (A) Male genitalia, lateral view; (B) Dorsal view of IXth segment; (C) Dorsal view of phallic apparatus (from Schmid, 1970).

Anal sclerite somewhat dumbbell shaped. Phallicata small and ventral lobe of phallicata flat with thick spines. Basal segment of inferior appendage broader at base and twice longer than terminal segment of inferior appendage. In ventral view terminal segment of inferior appendage with concavity.

Genitalia (♀): (Figs. 7.40A–B). VIIIth segment in a simple tube and truncated obliquely downwards. Copulatory bursa composed of three complex chitinous pieces, united by membranous folds.

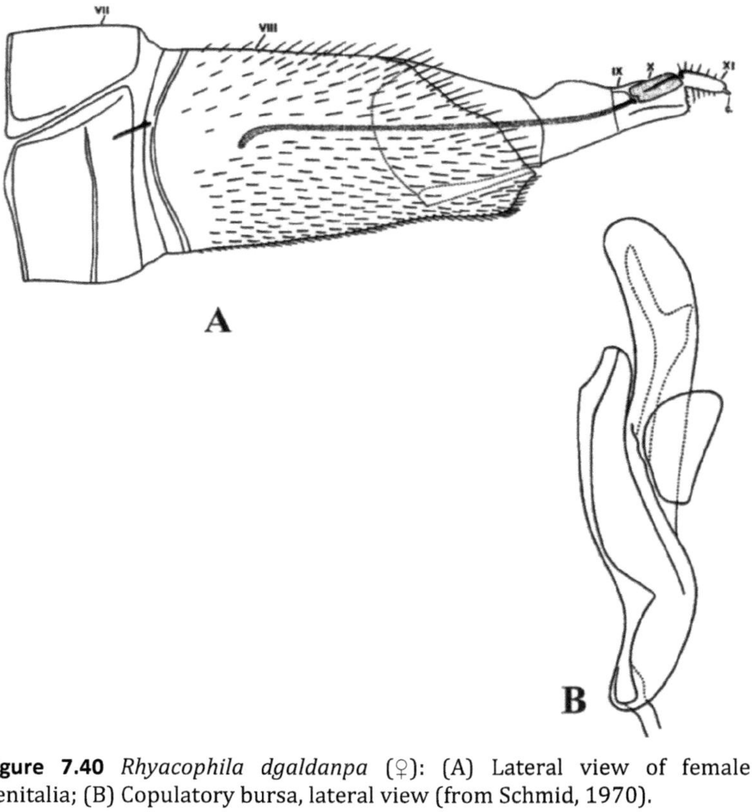

Figure 7.40 *Rhyacophila dgaldanpa* (♀): (A) Lateral view of female genitalia; (B) Copulatory bursa, lateral view (from Schmid, 1970).

Holotype: INDIA: Assam; NEFA, Kameng Frontier Division, Domkho, 11-V-1961.

Holotype depository: ROM, Canada.

Additional material examined: INDIA: Arunachal Pradesh: Menchuka, 2300 m, 29-IV-2010. (Coll. Parey & Pandhar).

Distribution: India (Arunachal Pradesh).

35. *Rhyacophila dikkaravasini* Schmid

Rhyacophila dikkaravasini, Schmid, 1970: 196

Adult: Upper body reddish-brown with dark golden hair. Yellowish leg, brown forewings, intensely speckled with golden macules, quite irregular in size and distribution. Wingspan: 19–22 mm.

Genitalia (♂): (Figs. 7.41A–B). Long and slender preanal appendages, with a very slight incision before their superior apical angle. Phallus long and slender; parameres like very fine spines, as long or a little longer than the phallus. Superior angle inferior appendages somewhat variable in shape, angular or rounded.

Genitalia (♀): Not available.

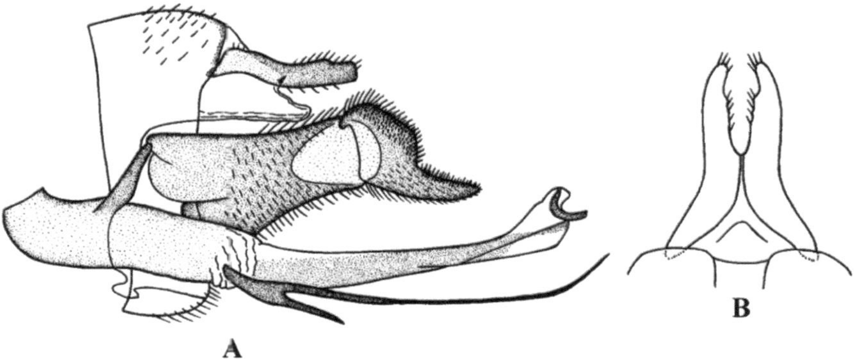

Figure 7.41 *Rhyacophila dikkaravasini* (♂): (A) Lateral view of male genitalia; (B) Dorsal view of preanal appendages (from Schmid, 1970).

Holotype: INDIA: Madras, Gudalur, 6500–7000 ft., 6-7-I-1959.

Holotype depository: ROM, Canada.

Distribution: India (Chennai).

36. ***Rhyacophila dilata*** **Martynov**

Rhyacophila dilata, Martynov, 1935: 96–97

Adult: Head, thorax, abdomen and wings brown; antenna with alternating light and dark brown bands; maxillary and labial palpi pale brown; legs orangish and forewing with dark brown irrotations. Length of antenna 7.76 mm and forewing 10.67 mm. In forewing fork 2nd arise earlier than 1st.

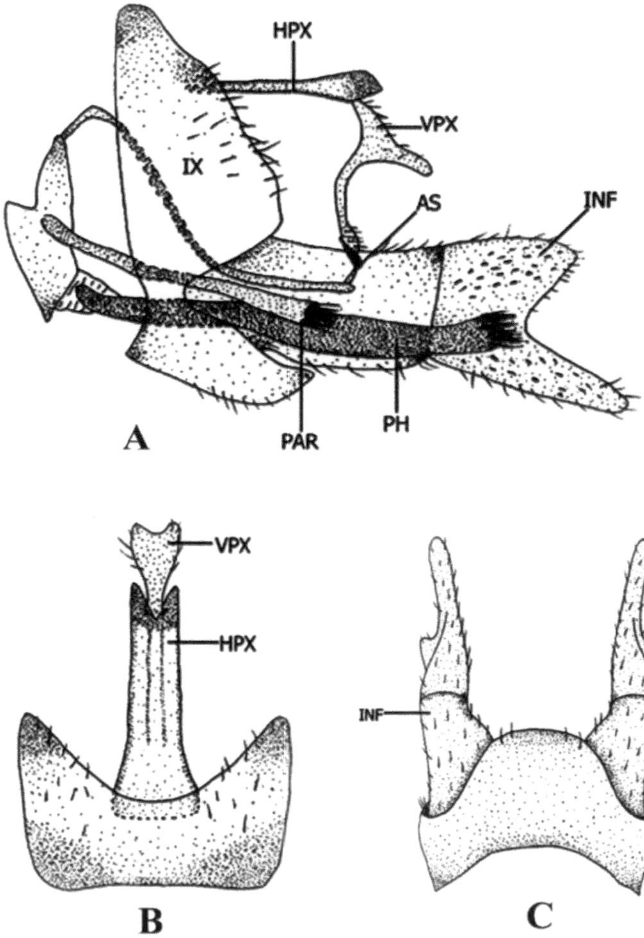

Figure 7.42 *Rhyacophila dilata* (♂): (A) Lateral view of male genitalia; (B) Dorsal view; (C) Ventral view.

Genitalia (♂): (Figs. 7.42A–C). In dorsal view superior point of vertical plate of segment X indented. Horizontal point of segment X in the form of an elongated plate, apex broader than base. In lateral view segment IX broad with deep, triangular excision near the inferior appendage; tergum broad, sternum forming a large oval projection under inferior appendage. Vertical plate of segment X shorter, directed downwards, rounded and setiferous. Inferior appendage large and very broad, basal segment comparatively short, at their bases covered by the projecting IX segment. Superior lobe of terminal segment of inferior appendage curved inwards, somewhat shorter than the inferior lobe. Paramere and phallicata with tuft of setae apically. In ventral view inferior lobe of inferior appendage rounded apically.

Genitalia (♀): Not available.

Holotype: INDIA: Phagu, 9000 ft., Shimla Hills; 18-21-v-1916.

Additional material examined: INDIA: Sikkim; Lachen, 3200 m, 24-V-2011. (Coll. Parey & Pandhar).

Distribution: Nepal, India (Himachal Pradesh, Uttarakhand, Sikkim).

37. *Rhyacophila dirangpa* Schmid

Rhyacophila dirangpa, Schmid, 1970: 199–200

Adult: Above of the body reddish yellow, with light golden hairs. Paws reddish yellow. Forewings brown, riddled with fine golden macules, evenly distributed. Wingspan: 17–22 mm.

Genitalia (♂): (Figs. 7.43A–C). Dorsal part of the Xth segment not very elongated and terminated in two hooks curved downwards. Dorsal appendix of the phallic apparatus bilobed, long and splited; parameres closely rounded at their end and armed with a row of lateral bristles. Inferior appendages obtuse, with its upper border widely and regularly rounded.

Genitalia (♀): Not available.

Holotype: INDIA: Assam; NEFA, Kameng Frontier Division, Rahung, 4500–8000 ft., 16-VIII-1961.

Holotype depository: ROM, Canada.

Distribution: India (Arunachal Pradesh).

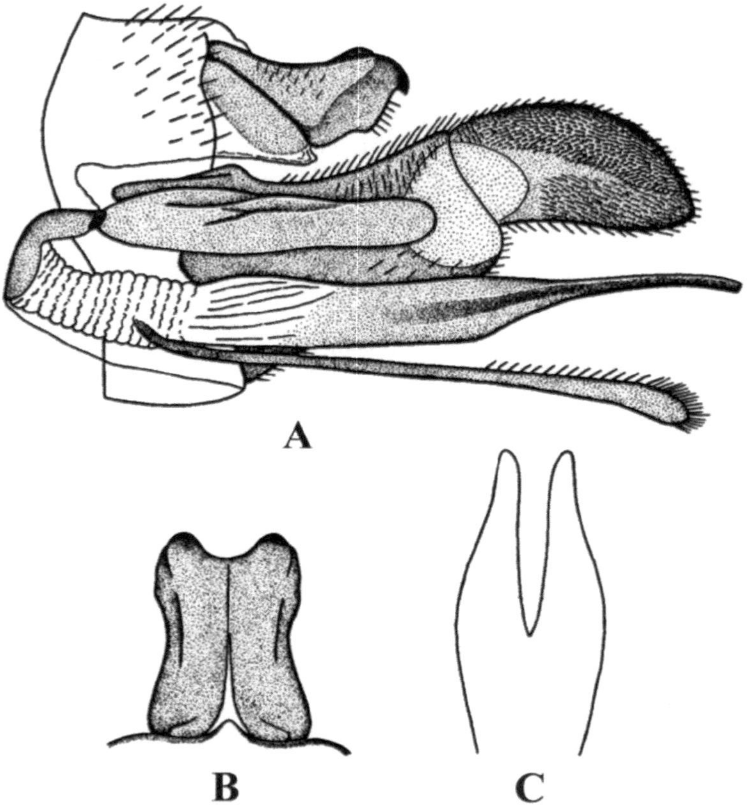

Figure 7.43 *Rhyacophila dirangpa* (♂): (A) Lateral view of male genitalia; (B) Dorsal view; (C) Phallic apparatus seen from above (from Schmid, 1970).

38. *Rhyacophila dolingpa* Schmid

Rhyacophila dolingpa, Schmid, 1970: 181

Adult: Upper part of the body reddish-brown, with brown hairs. Legs light reddish yellow, with mid and posterior showing orange coloration, evenly brown. Wingspan: 13–15.5 mm.

Genitalia (♂): (Figs. 7.44A–B). When seen laterally, dorsal lobe of the IXth segment appearing high and strongly convex. Preanal appendages and anal scleritis fully fused into an elongated and prominent piece. Phallus composed of two short sub-equal branches and markedly shorter than the dorsal appendages of the phallic apparatus. Inferior appendages regularly triangular.

Genitalia (♀): Not available.

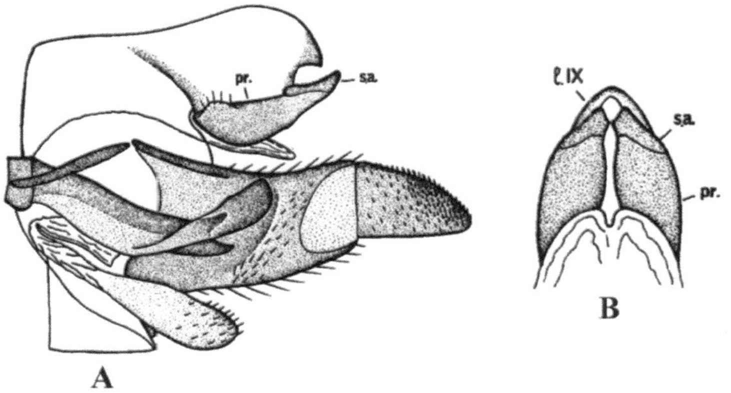

Figure 7.44 *Rhyacophila dolingpa* (♂): (A) Lateral view of male genitalia; (B) Dorsal view (from Schmid, 1970).

Holotype: INDIA: Assam; United Jaintia and Khasi Hills, Tarangblang, 300–4800 ft., 1-I-1960.

Holotype depository: ROM, Canada.

Distribution: India (Meghalaya).

39. *Rhyacophila dongkyapa* Schmid

Rhyacophila dongkyapa, Schmid, 1970: 138

Adult: Upper body dark brown, slightly pigmented and golden hairy. Legs usually light red, but sometimes brown. Forewings dark brown and riddled with small, evenly distributed golden macules; they are larger along the apical edge and quite extensive at the arculus. Wingspan: 13–20 mm.

Genitalia (♂): (Figs. 7.45A–D). End of the inner surface of the inferior appendages concave and with two longitudinal rows of strong spines; regular shaped, about three times as long as it is wide, and with a small upper basal prominence. Dorsal appendage of the phallic apparatus terminated in two rounded lobes; long parameres, sharply thinned and setiferous at their extremity.

Genitalia (♀): Not available.

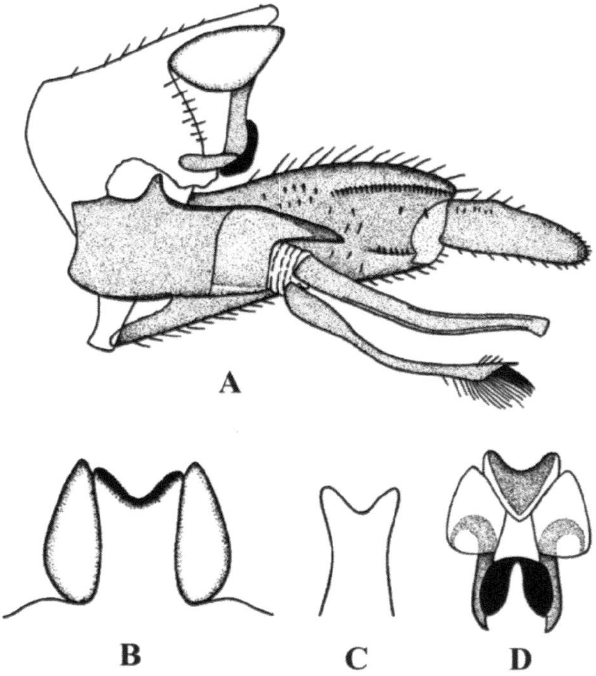

Figure 7.45 *Rhyacophila dongkyapa* (♂): (A) Male genitalia, lateral view; (B) Dorsal view; (C) Dorsal view of phallic apparatus; (D) Front view of Xth segment (from Schmid, 1970).

Holotype: INDIA: Assam; NEFA, Kameng Frontier Division, Brukpatarnchen, 1500–8000 ft., 17-III-1961.

Holotype depository: ROM, Canada.

Distribution: Bhutan, Nepal, India (Arunachal Pradesh).

40. *Rhyacophila dongre* Schmid

Rhyacophila dongre, Schmid, 1970: 191

Adult: Upper body dark reddish-brown, covered with brown and yellow hair. Reddish-brown paws. Forewings dark brown with concolorous hairiness, intermixed with golden hairs. Wingspan: 17–20 mm.

Genitalia (♂): (Figs. 7.46A–B). fairly high preanal appendages terminating in two small rounded and strongly divergent lobes. The lobes of anal sclerites also rounded, but much larger than the preceding ones. Inner section of Inferior appendages forming two lobes of unequal length and thickness and forming an acute angle between them. This species can be recognized by the markedly uneven apex of the preanal appendages and anal sclerites (Schmid, 1970).

Genitalia (♀): Not available.

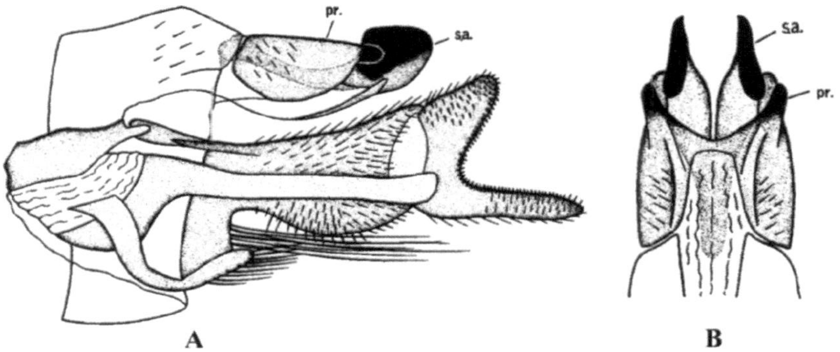

Figure 7.46 *Rhyacophila dongre* (♂): (A) Lateral view of male genitalia; (B) Dorsal view (from Schmid, 1970).

Holotype: INDIA: Sikkim; Yugang, 5100–5700 ft., 14-V-1959.

Holotype depository: ROM, Canada.

Distribution: India (Sikkim).

41. *Rhyacophila dorje* Schmid

Rhyacophila dorje, Schmid, 1970: 144

Adult: Upper body dark brown, with black hair. Dark reddish-brown legs. Forewings dark brown, with fine silvery macules, conspicuous. Wingspan: 13–15 mm.

Genitalia (♂): (Figs. 7.47A–D). In lateral view, the Xth segment appears similar to that of *R. lobsang,* although in front view, it does not form an X-shaped figure, but a chitinous band, simple vertical. Inferior appendages not or very weakly notched. Rather short phallic apparatus; phallus usually small, but sometimes as thick as that of *R. lobsang;* ventral plate with a terminal scaffold of variable shape, but always large (Schmid, 1970).

Genitalia (♀): Not available.

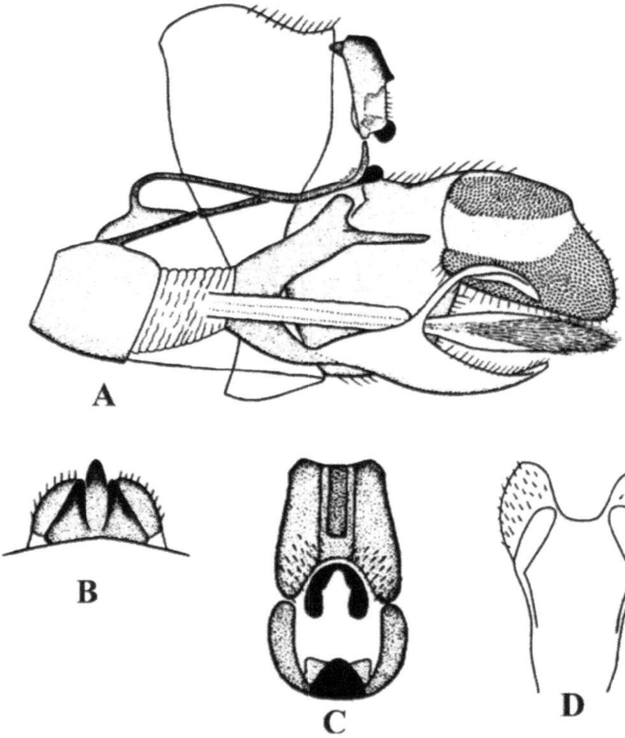

Figure 7.47 *Rhyacophila dorje* (♂): (A) Male genitalia, lateral view; (B) Dorsal view of Xth segment; (C) Front view of Xth segment; (D) Ventral plate of the phallus, seen from above (from Schmid, 1970).

Holotype: INDIA: Madras (Chennai), Perumalmalai, 1500–7000 ft., 3-6-XII-1958.

Holotype depository: ROM, Canada.

Distribution: India (Chennai).

42. *Rhyacophila drokpa drokpa* Schmid

Rhyacophila drokpa drokpa, Schmid, 1970: 220–221

Subspecies: *Rhyacophila drokpa gurla,* Schmid, 1970

Subspecies: *Rhyacophila drokpa nyenpa,* Schmid, 1970

Adult: Entire body brown and forewing with dark brown irrotations. Length of antenna 8.73 mm and forewing 11.0 mm. In forewing fork 1st arise slightly earlier than 2nd.

Genitalia (♂): (Figs. 7.48A–C). In dorsal view anal sclerite and dorsal apical lobe of segment IX rounded apically. Segment X sub-truncate in shape. In lateral view, segment IX elongated and with well-developed dorsal apical lobe of segment IX. Segment X broader apically and dorsal appendage of phallic apparatus slightly extending beyond anal sclerite. Phallicata small and ventral lobe of phallicata bulbous and setiferous. Basal segment of inferior appendage longer than terminal segment of inferior appendage. Superior lobe of terminal segment of inferior appendage smaller than the inferior lobe. Inferior lobe finger like. In ventral view terminal segment of inferior appendage rounded.

Genitalia (♀): Not available.

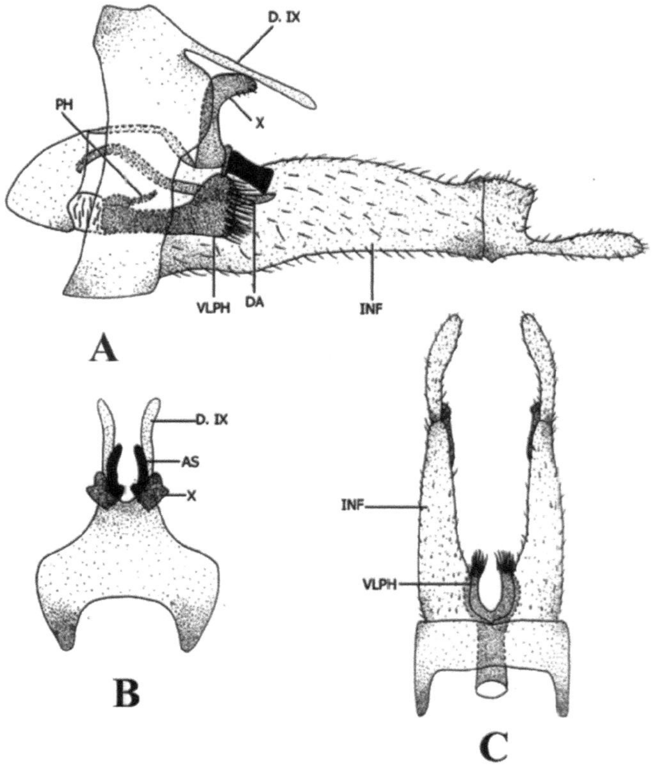

Figure 7.48 *Rhyacophila drokpa drokpa* (♂): (A) Lateral view of male genitalia; (B) Dorsal view; (C) Ventral view.

Holotype: INDIA: Assam; NEFA, Kameng Frontier Division, Kujjalong, 5300–8000 ft., 28-30-VI-1961.

Holotype depository: ROM, Canada.

Distribution: Bhutan, Thailand, India (Arunachal Pradesh, Uttarakhand, Assam).

43. *Rhyacophila drosampa* Schmid

Rhyacophila drosampa, Schmid, 1970: 162–163

Adult: Upper body black, with dark brown hair. Dark brown underside and legs. Wings uniformly black. wingspan: 10–25 mm.

Genitalia (♂): (Figs. 7.49A–B). Dorsal piece appearing to be long oval and pointed, viewed from above, with the suture between the lobe of the IXth segment and the preanal appendages clearly visible. Small anal scleritis. Superior apical angle of the inferior appendages pointed; broadly rounded lower angle.

Genitalia (♀): Not available.

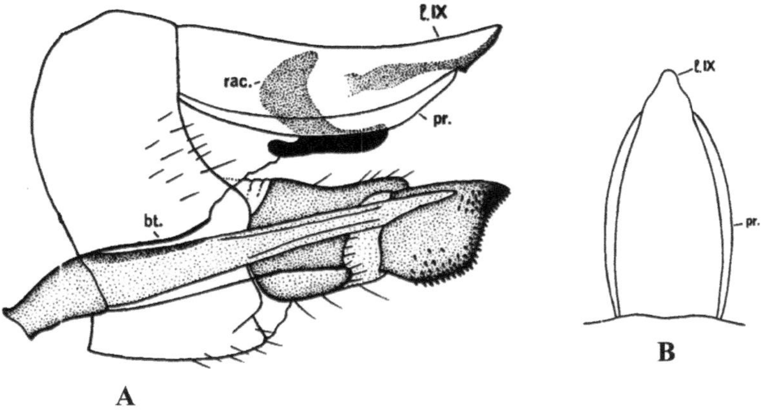

Figure 7.49 *Rhyacophila drosampa* (♂): (A) Lateral view of male genitalia; (B) Dorsal view (from Schmid, 1970).

Holotype: INDIA: Assam (Now Meghalaya), United Jaintia and Khasi Hills, Rumkheng, 20-23-III-1960.

Holotype depository: ROM, Canada.

Distribution: Thailand, India (Meghalaya).

44. *Rhyacophila drotangpa* Schmid

Rhyacophila drotangpa, Schmid, 1970: 138–139

Adult: Above of the body blackish, with golden hair. Claws of the three pairs of legs larger than usual. Remarkably long and narrow wings; the anterior ones are uniformly golden and the posterior ones gray, but the fringes are dark gray and the pterostigma well marked on both wings. Wingspan: 18.5 mm.

Genitalia (♂): (Figs. 7.50A–D). Apical dorsal lobe of the IXth segment with a rounded notch and terminated in two thin lateral branches. Inferior appendages robust, with the second segment triangular and very elongated. Dorsal appendix of phallic apparatus in narrow oval; parameres simple and slender, with fine apical bristles. This species is closely related to *R. yigrongpa*, but the lower appendages are more robust and the abdomen dark (Schmid, 1970).

Genitalia (♀): Not available.

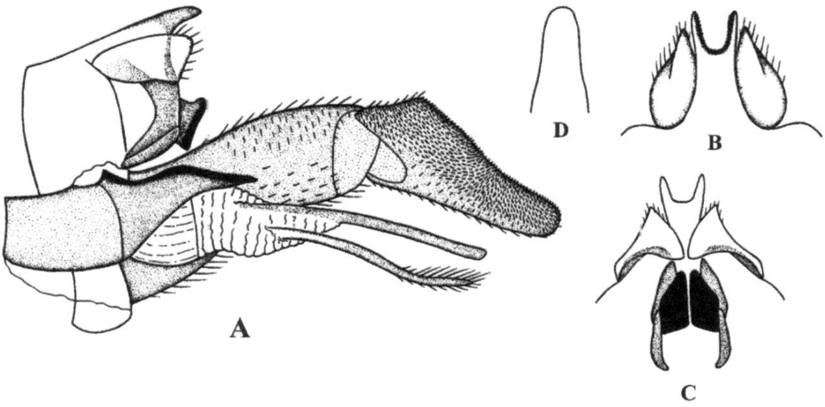

Figure 7.50 *Rhyacophila drotangpa* (♂): (A) Lateral view of male genitalia; (B) Dorsal view; (C) Front view of Xth segment; (D) Apex of the phallus (dorsal) (from Schmid, 1970).

Holotype: INDIA: Sikkim; Karponang, 9,900 ft., 23-VIII-1959.

Holotype depository: ROM, Canada.

Distribution: India (Sikkim).

45. *Rhyacophila extensa* Martynov

Rhyacophila extensa, Martynov, 1927: 493

Syn: *R. carletoni* Banks, 1931

Adult: Dorsal side of head brownish, yellowish behind, with yellowish hairs; ocelli large and yellow; joints of antennae yellow. Thorax brown above, sides brownish; legs yellow. Abdomen brown; basal half yellow beneath. Anterior wings elongated, apex distinctly stretched; coloration pale, with greyish-brownish transverse bands, becoming indistinct in the middle portion; provided with rather large hyalin irrorations; behind the Cu, about 2–3 pale brownish spots. Apical fork first beginning somewhat earlier, than the second one; in the posterior wing the first fork commences also earlier than the second.

Genitalia (♂): (Figs. 7.51A–B). IXth segment yellow; median process of the IXth tergite very long, reaching the end of the basal joint of the inferior appendage, somewhat narrowing to the apex. Basal joint of the inferior appendage long, dilating to the end (from side), as well as to the base; second joint is also rather long; upper lobe very short, and represented by only an angle, lower lobe slender, digitiform. Phallus long and slender. Length of the body, with genital append ages, 7.7 mm.

Genitalia (♀): Not available.

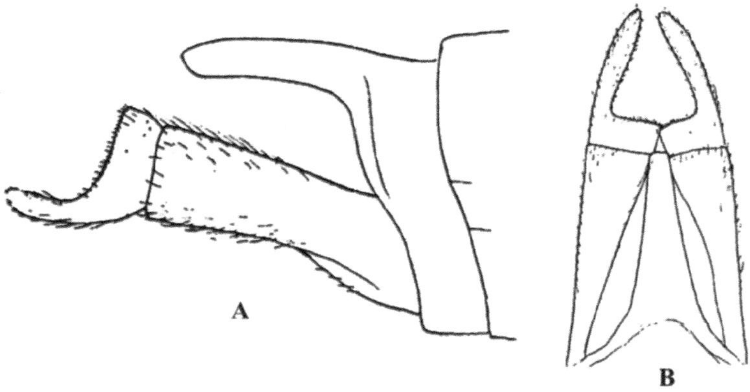

Figure 7.51 *Rhyacophila extensa* (♂): (A) Lateral view of male genitalia; (B) Dorsal view (from Martynov, 1927).

Holotype: TURKESTAN: Przewalsk district, Semiretchje, 1920.

Holotype depository: Ann. Mus. Zool. Ac. Sci., USSR.

Distribution: Turkestan, India (Himachal Pradesh).

46. *Rhyacophila fletcheri* (Kimmins)

Rhyacophila fletcheri (Kimmins), Ross, 1956: 115 (New combination).

Syn: *Trichophila fletcheri* Kimmins, 1952: 359–361

Adult: Entire body brown except hind legs which are pale brown and antenna with alternating light and dark brown bands. Length of antenna 7.76 mm and forewing 8.73 mm. In forewing fork 1st arise earlier than 2nd.

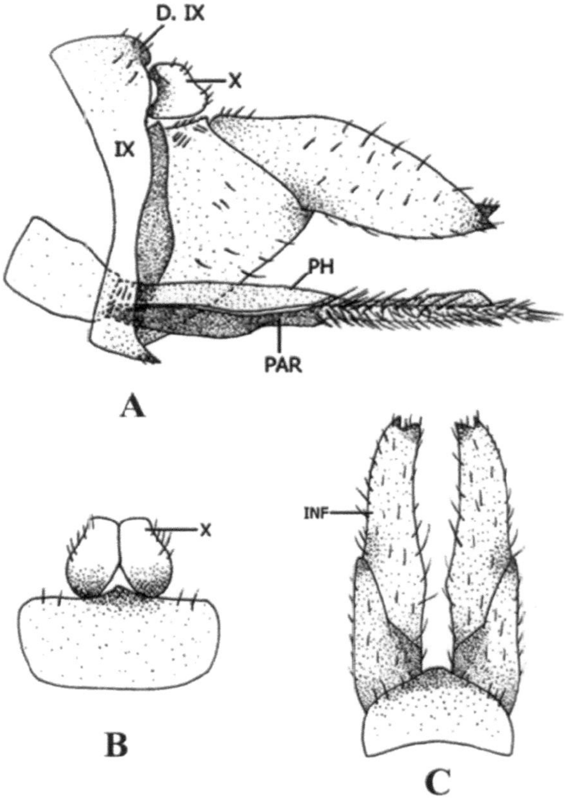

Figure 7.52 *Rhyacophila fletcheri* (♂): (A) Lateral view of male genitalia; (B) Dorsal view; (C) Ventral view.

Genitalia (♂): (Figs. 7.52A–C). In dorsal view segment X composed of two fused lobes, rounded from base. In lateral view segment IX with slightly produced dorsal apical lobe of segment IX. Paramere bulbous at base and densely clothed with stout setae apically. Phallicata slender shaped. Inferior appendage two-segmented, basal segment broader than the terminal segment. Terminal segment large, irregular ovate. Tergum and posterior margin of segment IX rounded whereas posterior margin of sternum is pointed. In ventral view terminal segment of inferior appendage with V shape indentation. *R. fletcheri* (Kimmins, 1952) almost different from rest members of genus *Rhyacophila* due to the morphology of inferior appendage, segment X, phallicata and paramere.

Genitalia (♀): Not available.

Holotype: INDIA: Sikkim; Kurseong, 5000 ft., 18-30-IV-1922.

Holotype depository: BMNH, London.

Additional material examined: INDIA: Uttarakhand; Badrinath, 3400 m, 22-VI-2011, 2 ♂♂. (Coll. Parey & Parey).

Distribution: India (Arunachal Pradesh, Sikkim, Uttarakhand).

47. *Rhyacophila gelukpa* Schmid

Rhyacophila gelukpa Schmid, 1970: 214–215

Adult: Entire body pale brown except head and mesothorax which are brown. Length of antenna 9.7 mm and forewing 11.64 mm. In forewing fork 2nd arise slightly earlier than 1st.

Genitalia (♂): (Figs. 7.53A–C). In dorsal view apex of dorsal apical lobe of segment IX less broad than base. In lateral view dorsal apical lobe of segment IX triangular shaped apically. Anal sclerite slightly hidden beneath segment X. Dorsal appendage of phallic apparatus rounded apically. Phallicata and paramere almost of same length and paramere bifid and upper part of paramere is longer than the lower one. Terminal segment of inferior appendage with U shape indentation. In ventral view inferior and superior lobes of terminal segment of inferior appendage rounded apically.

Genitalia (♀): Not available.

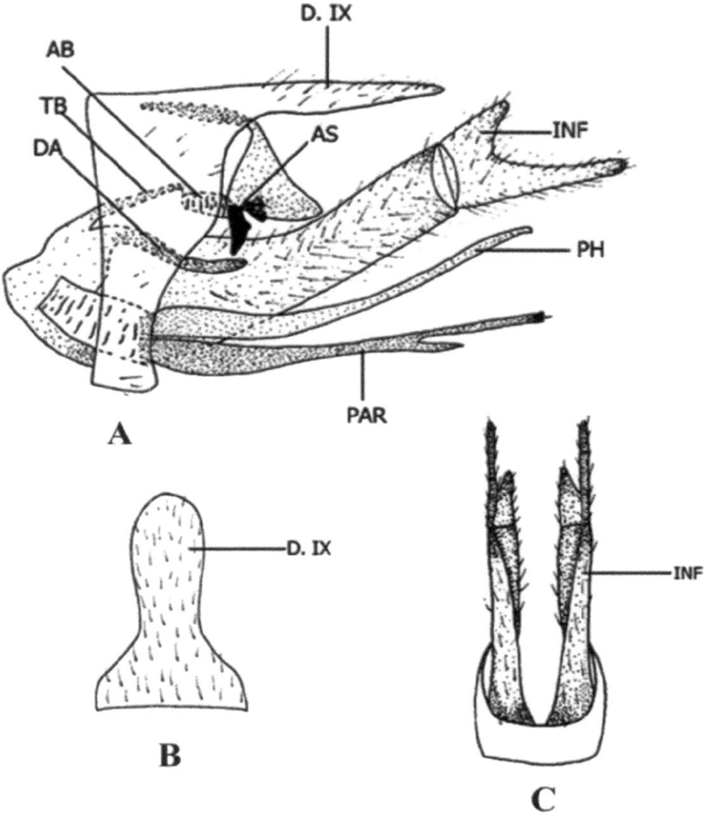

Figure 7.53 *Rhyacophila gelukpa* (♂): (A) Lateral view of male genitalia; (B) Dorsal view; (C) Ventral view.

Holotype: INDIA: Sikkim; Gey, 5300–9000 ft., 18-V-1959.

Holotype depository: ROM, Canada.

Additional material examined: INDIA: Arunachal Pradesh; Bomdila, 2400 m, 03-X-2010, 4 ♂♂; Sikkim; Lachen, 3200 m, 24-V-2011, 1 ♂. (Coll. Parey & Pandhar).

Distribution: Vietnam, India (Sikkim, Arunachal Pradesh).

Population Variation: (Figs. 7.54A–C). The specimens of *R. gelukpa* from Bomdila (Andhra Pradesh) show variations from specimen of same species from Lachen (Sikkim). From Bomdila the individuals of *R. gelukpa* are with black spine structure on place of furcation of paramere whereas this structure is absent in individual from Lachen. Individuals from both localities also show difference in shape of anal sclerite and shape of upper part of paramere apically. Dorsal apical lobe of IXth segment in dorsal as well as in lateral view also show variation in individuals from Lachen and Bomdila. In individuals from Bomdila, in dorsal view, dorsal apical lobe of IXth segment with indent and in lateral view dorsal apical lobe of IXth segment rounded apically however individual from Lachen, in dorsal view, dorsal apical lobe of IXth segment without any indent and in lateral view dorsal apical lobe of segment IX triangular shaped apically.

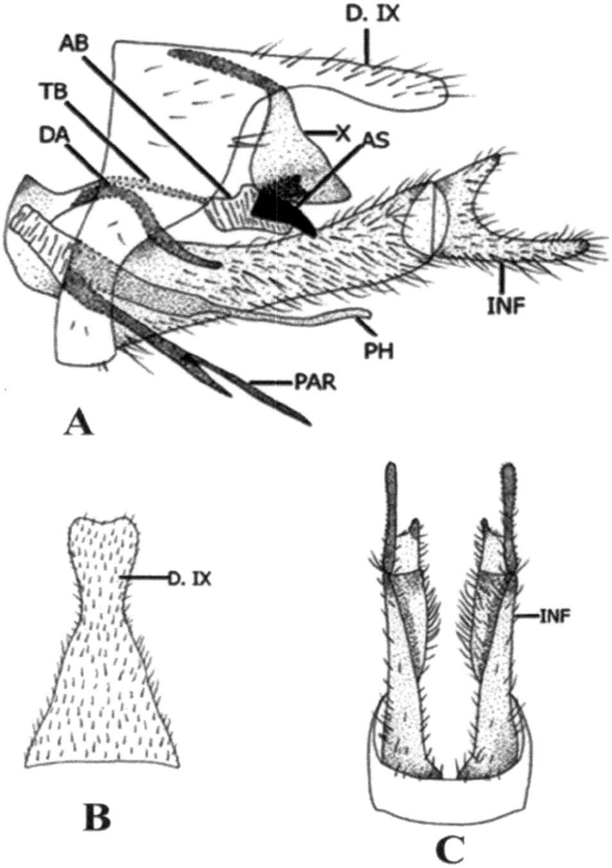

Figure 7.54 *Rhyacophila gelukpa* (Variant) (♂): (A) Lateral view of male genitalia; (B) Dorsal view; (C) Ventral view.

48. *Rhyacophila gyaldzen* Schmid

Rhyacophila gyaldzen Schmid, 1970: 144–145

Adult: Abcve of the body very dark reddish-brown, with concolorous hairiness; dark red paws. Forewings very dark and riddled with conspicuous silvery macules. Wingspan: 13–14.5 mm.

Genitalia (♂): (Figs. 7.55A–D). Upper part of the Xth segment developed backwards and armed with two short subparallel teeth directed downwards. Phallus very robust and a little curved in S. Inferior appendages small, simple in shape and slightly indented.

Genitalia (♀): Not available.

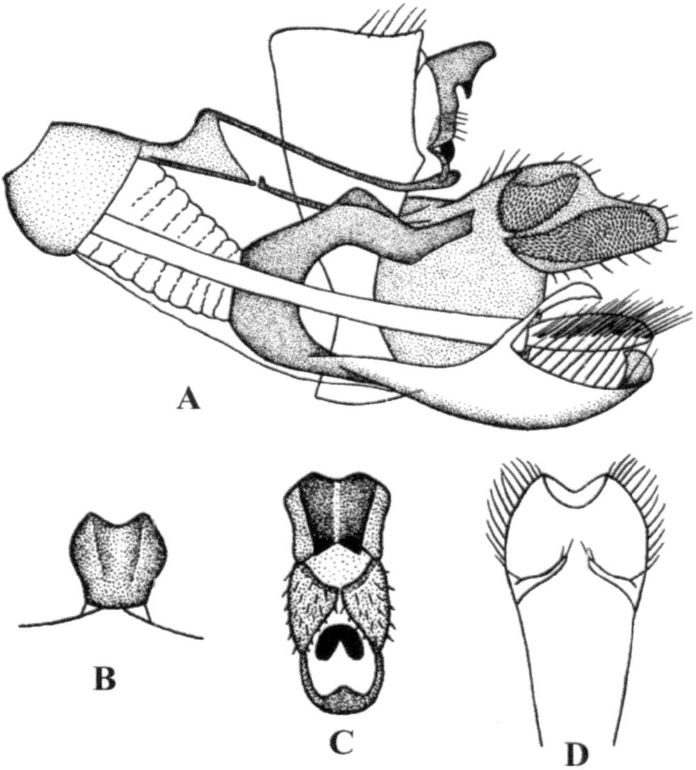

Figure 7.55 *Rhyacophila gyaldzen* (♂): (A) Lateral view of male genitalia; (B) Dorsal view of Xth segment; (C) Front view of Xth segment; (D) Ventral plate of phallus (Dorsal view) (from Schmid, 1970).

Holotype: INDIA: Assam (Now Meghalaya); United Jaintia and Khasi Hills, Muktapur 11-12-XII-1959.

Holotype depository: ROM, Canada.

Distribution: India (Meghalaya).

49. *Rhyacophila gyamo* Schmid

Rhyacophila gyamo Schmid, 1970: 180–181

Adult: Above of the body black-brown, with black hair. Upper part of the pleura and legs reddish-brown, but the lower part of the median and posterior pleura and the corresponding femora and tibia are bright orange-yellow. Wings uniformly black. Wingspan 9: 15–19 mm.

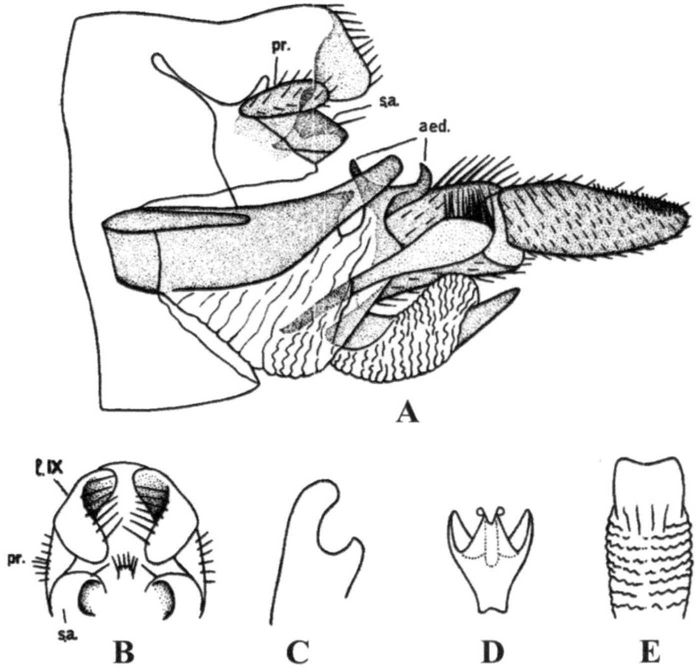

Figure 7.56 *Rhyacophila gyamo* (♂): (A) Lateral view of male genitalia; (B) Dorsal assembly, seen from face; (C) Dorsal appendage of the phallic apparatus, seen from above; (D) Apex of the phallus; (E) Ventral plate of the phallus, seen from above (from Schmid, 1970).

Genitalia (♂): (Figs. 7.56A–E). Preanal appendages in small ovals. Anal sclerites in obtuse triangles. Ventral lobe of the membranous phallus and terminated by a thick, rectangular chitinous plate. Inferior appendages suboval not indented and two and a half times longer than high.

Genitalia (♀): Not available.

Holotype: INDIA: Manipur; Sirohi Kashong, 5400–8000 ft., 6-7-VI-1960.

Holotype depository: ROM, Canada.

Distribution: Thailand, India (Manipur).

50. *Rhyacophila gyaspa* Schmid

Rhyacophila gyaspa Schmid, 1970: 200

Adult: Upper body light reddish-brown, with golden hairs. Pleura and paws reddish yellow. Fairly wide wings: the anterior ones are light golden brown, densely, evenly, and finely spotted with dark, but there are two narrow dark transverse bands on the anastomosis and in the middle of the apical cells. Wingspan: 17–28 mm.

Genitalia (♂): (Figs. 7.57A–C). When viewed from above, the dorsal part of the Xth segment appears broad, its external apical angles in triangular fins and with a clear apical median notch. Dorsal appendage of the phallic apparatus not very long; Parameres widened at the apex and with a single dense bouquet of thick preapical spines. Inferior appendage rather slender and with two terminal lobes of very unequal sizes.

Genitalia (♀): Not available.

Holotype: INDIA: Western Bengal; Dilpa, 2300–5800 ft., 2-III-1959.

Holotype depository: ROM, Canada.

Distribution: India (West Bengal).

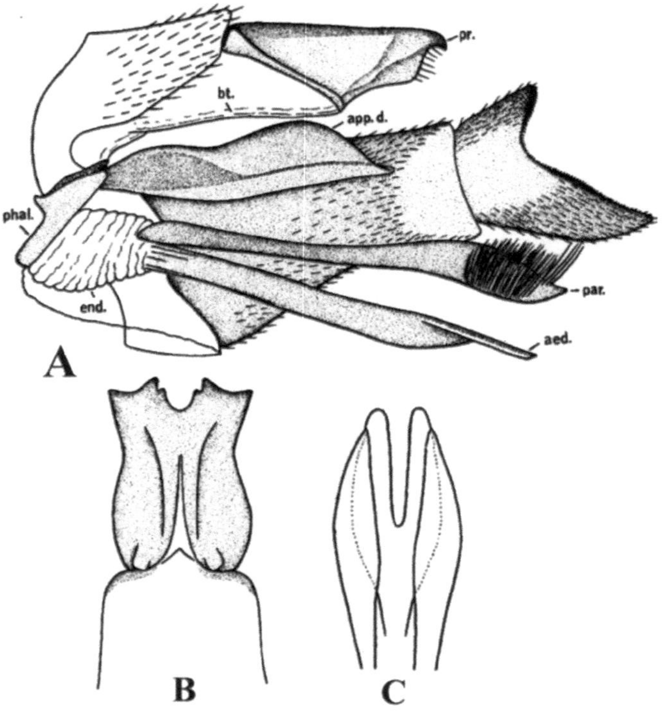

Figure 7.57 *Rhyacophila gyaspa* (♂): (A) Lateral view of male genitalia; (B) Dorsal view; (C) Dorsal appendage of the phallic apparatus, seen from above (from Schmid, 1970).

51. *Rhyacophila gyelbu* Schmid

Rhyacophila gyelbu Schmid, 1970: 187

Adult: Upper body dark reddish-brown, with concolorous hairiness. Pleura and paws reddish yellow. Wings uniformly dark brown. Wingspan: 14–19 mm.

Genitalia (♂): (Figs. 7.58A–B). Apical dorsal lobe of IXth segment very briefly incised at its end. Anal sclerites with their external border strongly raised in a high very oblique parallelogram and with the projecting internal apical angle. Phallus large and composed of two very unequal branches; missing parameres. Inferior appendages slightly incised at the base of their upper edge.

Genitalia (♀): Not available.

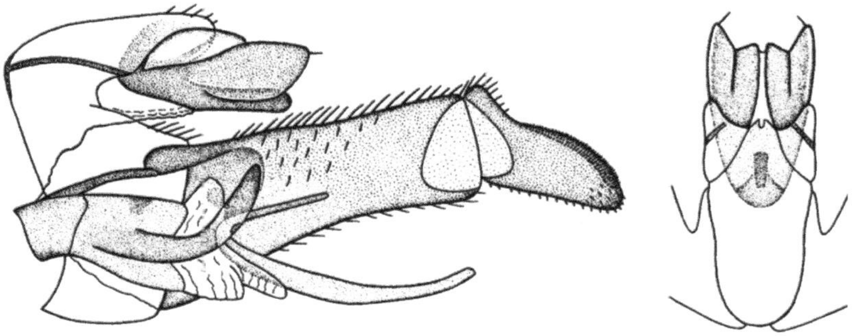

Figure 7.58 *Rhyacophila gyelbu* (♂): (A) Lateral view of male genitalia; (B) Dorsal view (from Schmid, 1970).

Holotype: INDIA: Manipur; Sirohi Kashong, 6800–8500 ft., 6-7-VI-1960.

Holotype depository: ROM, Canada.

Distribution: India (Manipur).

52. *Rhyacophila himachalensis* Kaur and Saini

Rhyacophila himachalensis Kaur and Saini, 2012: 212

Adult: Head, antenna and abdomen dark brown. Maxillary and labial palpi pale brown. Prothorax pale brown but meso and metathorax dark brown in color. Foreleg, midleg and hindleg light brown but tarsi of foreleg and hindleg dark brown. Length of antenna 4.85 mm and forewing 7.76 mm and both wings light brown but dark brown spot on costal margin of both the wings. In forewing fork 2nd arise slightly earlier than 1st.

Genitalia (♂): (Figs. 7.59A–E). Dorsal apical lobe of segment IX with indentation and preanal appendage oblong in dorsal view. In lateral view, preanal appendage longer than dorsal apical lobe of segment IX and dorsal appendage of phallic apparatus longer than the paramere and rounded at apex and heavily sclerotized. The phallicata is tubular and paramere with tuft of thick setae apically. The dorsal edge of terminal segment of inferior appendage convex and basal segment of inferior appendage three times longer than the terminal segment. Stout spines present on the basal segment of inferior appendage. In ventral view terminal segment of inferior appendage with thick spines.

Genitalia (♀): Not available.

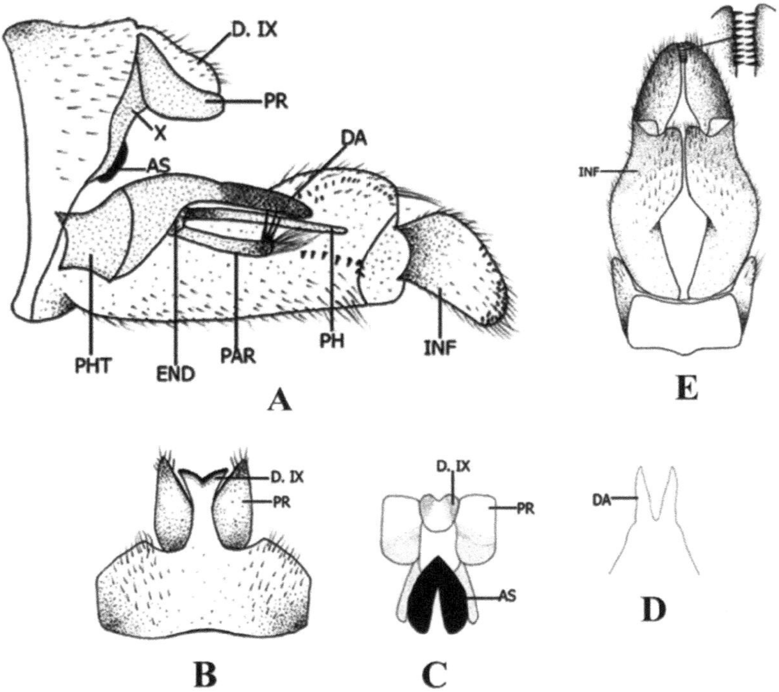

Figure 7.59 *Rhyacophila himachalensis* (♂): (A) Male genitalia, lateral view; (B) Dorsal view; (C) Caudal view; (D) Dorsal view of dorsal appendage of Phallicata; (E) Ventral view.

Holotype: INDIA: Himachal Pradesh; Banjar, 950 m, 30-VI-2009. (Coll. Parey & Pandhar).

Holotype depository: MDZPUP, India.

Distribution: India (Himachal Pradesh, Uttarakhand).

53. *Rhyacophila himalayensis* Saini, Kaur and Bajwa

Rhyacophila himalayensis Saini, Kaur and Bajwa, 2013: 224

Adult: Body color uniformly fuscous, pleura and legs light red; forewing sprinkled with small golden spots. Body entirely covered with inconspicuous golden pubescence; forewing uniformly covered with fuscous pubescence. Average length of forewing 5.5 mm.

Genitalia (♂): (Figs. 7.60A–C). Laterally, dorsal apical lobe of IXth segment curved downwards, preanal appendage oblong and anal sclerite extending beyond dorsal apical lobe of segment IX. Basal segment of inferior appendage twice as broad at the base as the terminal segment. Dorsal appendage of phallic apparatus folded apically. Ventral lobe of Phallicata bulbous and longer than the other parts of phallic apparatus.

Genitalia (♀): Not available.

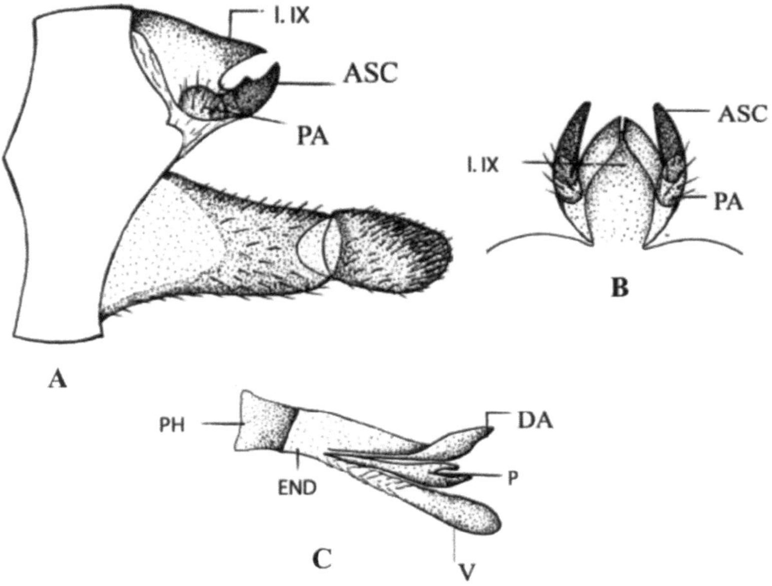

Figure 7.60 *Rhyacophila himalayensis* (♂): (A) Lateral view of male genitalia; (B) Dorsal view; (C) Phallic apparatus.

Holotype: INDIA: Uttarakhand; Badrinath, 3096 m. 28-VI-2000. (Coll. Parey & Pandhar).

Holotype depository: Museum of Department of Zoology & Environmental Sciences, Punjabi University, Patiala, India.

Distribution: India (Uttarakhand).

54. *Rhyacophila hingstoni* Martynov

Rhyacophila hingstoni Martynov, 1930: 67–69

Adult: Head, thorax, wings and abdomen brown; antenna with dark and pale brown alternate bands; maxillary and labial palpi pale brown; forewing with dark brown irrotations; foreleg and midleg with distinct markings; hindleg pale brown. Length of antenna 14.5 mm and forewing 16.0 mm. In forewing fork 2nd arise slightly earlier than 1st.

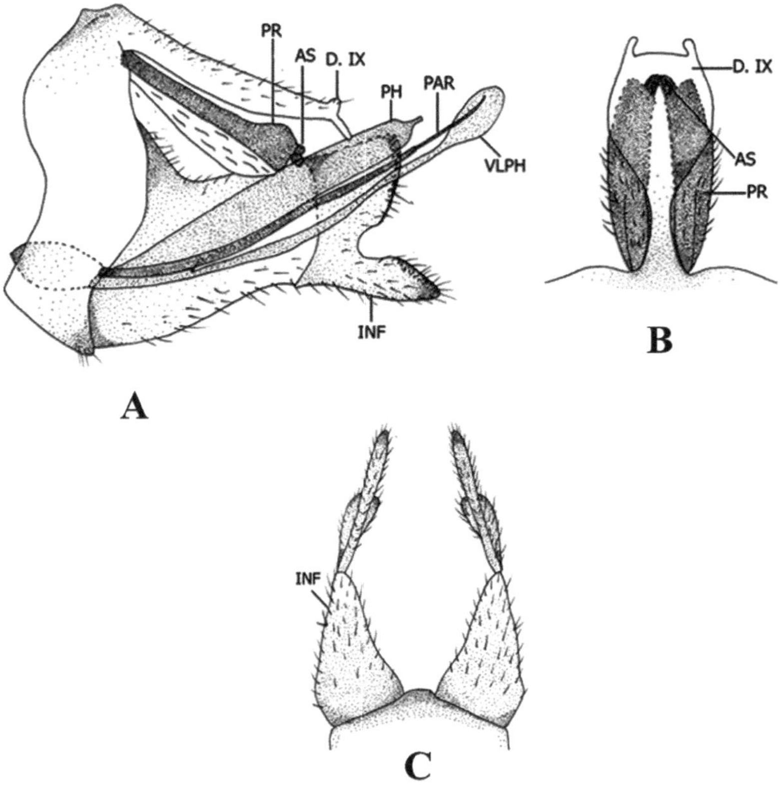

Figure 7.61 *Rhyacophila hingstoni* (♂): (A) Lateral view of male genitalia; (B) Dorsal view; (C) Ventral.

Genitalia (♂): (Figs. 7.61A–C). In dorsal view anal sclerite and some part of preanal appendage hidden below dorsal apical lobe of segment IX. In lateral view segment IX has its upper surface excised and from center of excision extends long plate-like dorsal process, this dorsal apical lobe of segment IX broader at base and then tapered at apex. Inferior appendage long, basal segment twice as long as terminal segment; terminal segment with rounded excision in lower half of apical margin, forming quadrate superior lobe, longer and digitate inferior lobe. Preanal appendage hinged to base of the dorsal apical lobe of segment IX. Anal sclerite bilobed. Phallicata long, laterally compressed, apex from laterally produced in small finger, from lower surface of phallicata arises ventral lobe, which is slender and extending beyond apex of phallicata, with flattened clavate apex. Paramere as long as ventral lobe of Phallicata. In ventral view inferior lobe of terminal segment of inferior appendage twice longer than superior lobe.

Genitalia (♀): Not available.

Holotype: INDIA: Sikkim; Khamba Jong, 15000–16000 ft., 15-30-VII-1903.

Holotype depository: British Museum, London.

Additional material examined: INDIA: Sikkim; Lachung, 2600 m, 16-IX-2009. (Coll. Parey & Pandhar).

Distribution: India (Uttarakhand).

55. *Rhyacophila hobsoni* Martynov

Rhyacophila hobsoni Martynov, 1930: 69

Adult: Head yellowish, with brown warts. Basal joint of antennae brownish. Thorax brownish, paler beneath. Legs yellowish, with usual fuscous markings on anterior and median pairs. Anterior wings somewhat elongate, resembling those of *R. hingstoni*; at terminations of nervures similar brownish spots, but less distinct; in apical portion of wing two short transverse brownish areas between Cu and A-l rather pale; forks 1 and 2 beginning at the same level, or 2nd fork extending further inward than 1st. Abdomen brown dorsally, pale beneath.

Genitalia (♂): (Figs. 7.62A–B). IXth segment widely excised above, the center of the excision produced in a long dorsal process, tapering to about midway and then dilating slightly to a shallowly excised apex, lateral angles a little produced downward. In side view, X^{th} tergite similar in construction to *R. hingstoni*, but the upper apical angle is rectangularly excised and the bilobed structure is relatively larger. X^{th} sternite membranous. Phallus long, laterally compressed, straight. The upper margin, in side view terminates in a pair of small, acute spines set side by side, and the lower margin extends in a strong sinuous hook. Beneath the phallus arises a pair of spiniform processes, broad at their bases, tapering to sub-acute at apex, nearly as long as the phallus. Inferior appendage rather longer; basal segment less deep, terminal segment with a wider apical excision, upper lobe rounded and wider than lower.

Genitalia (♀): Not available.

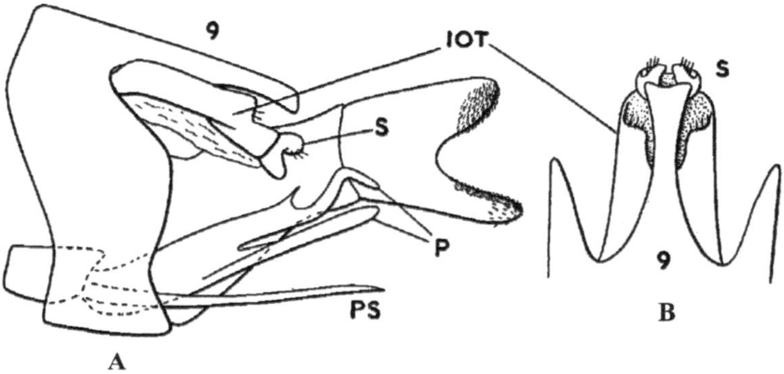

Figure 7.62 *Rhyacophila hobsoni* (♂): (A) Male genitalia, lateral view; (B) Dorsal view of IXth and X^{th} tergites (from Kimmins 1953).

Holotype: TIBET: Yatung, 15000–16000 ft., 15-30-VI-1903.

Holotype depository: British Museum, London.

Distribution: Tibet, Nepal, India (Sikkim, Uttarakhand).

56. *Rhyacophila inconspicua* Morton

Rhyacophila inconspicua Morton, 1900: 7

Adult: The head and thorax are black-brown in both sexes. Wings elongate, apex elliptical, apical fork 2 extending further inwards than I. Radius not distinctly furcate. Costal hairs simple. Membrane of wings smoky, clothed with brownish pubescence, fringes dark gray, neuration fuscous hairs of vertex black; legs fuscescent.

Genitalia (♂): (Figs. 7.63A–B). The oval dorsal process is divided by a median line; at its apex appear two approximated processes which gradually narrow to their acute points which are divergent; at either side of the dorsal process is a small somewhat oval piece, and beneath is apparently a pair of blades obliquely truncate on apical edge. The inferior appendages have the first joint moderately long, the upper and lower edges parallel; second joint short on the upper margin, the lower margin twice as long, apical edge oblique, very slightly concave in outline, the inner side of this joint being thickly set with very short setae.

Genitalia (♀): Not available.

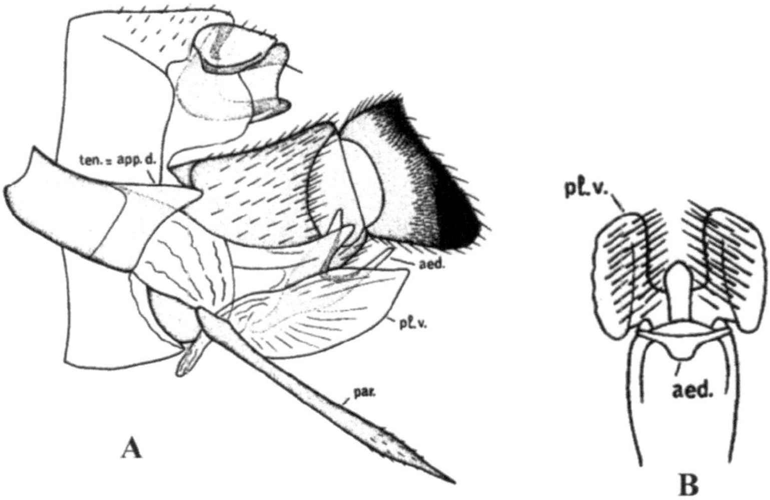

Figure 7.63 *Rhyacophila inconspicua* (♂): (A) Male genitalia, lateral view; (B) Dorsal view of phallic apparatus (from Schmid, 1970).

Holotype: INDIA: Assam; Khasi Hills; 6-XII-1899.

Holotype depository: BMNH, London.

Distribution: India (Meghalaya).

57. *Rhyacophila indica* Saini, Kaur and Bajwa

Rhyacophila indica Saini, Kaur and Bajwa, 2012: 224–226

Adult: Body uniformly fuscous ferruginous, pleura yellowish, forewing dark fuscous and intensely spotted with golden coloration. Body densely covered with inconspicuous bicolored pubescence. Setae on head wart as long as the scape. Length of forewing 4.5 mm.

Genitalia (♂): (Figs. 7.64A–C). Dorsal apical lobe of IXth segment indented apically and shorter than the preanal appendage. From above, preanal appendage constricted at apical half. Inferior appendage long, first segment bearing single row of spines on internal surface, three times longer than terminal segment. X^{th} segment slanting/oblique. Anal sclerite closer to segment X. Phallic apparatus with dorsal appendage chitinised apically; phallicata longer than paramere and tapering apically; paramere with tuft of setae at the apex.

Genitalia (♀): Not available.

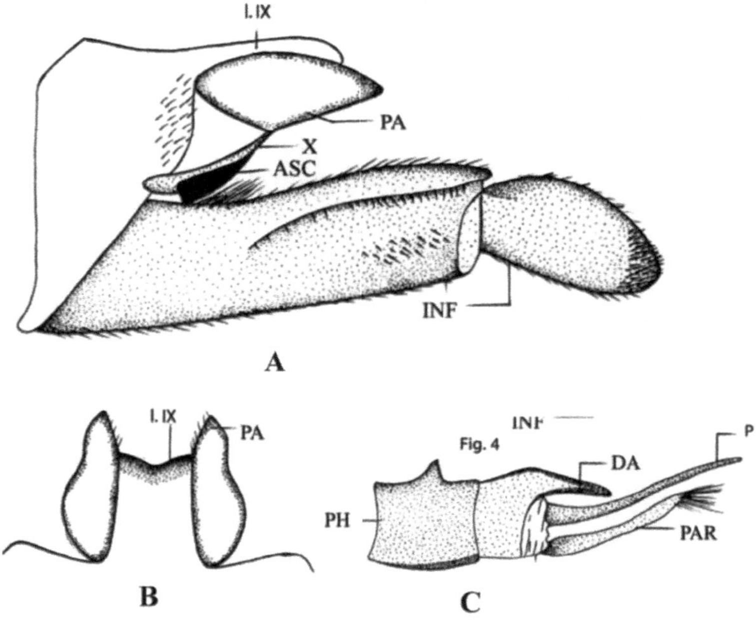

Figure 7.64 *Rhyacophila indica* (♂): (A) Lateral view of male genitalia; (B) Dorsal view; (C) Phallic apparatus.

Holotype: INDIA: Sikkim; Gangtok, 1700 m, 25-V-1999.

Holotype depository: Museum of Department of Zoology & Environmental Sciences, Punjabi University, Patiala, India.

Distribution: India (Sikkim).

58. *Rhyacophila jayadurga* Schmid

Rhyacophila jayadurga, Schmid, 1970: 196

Adult: Above of the body dark reddish-brown, with abundant concolorous hairiness. Pleura and legs light reddish yellow. Golden forewings, dotted with brown spots, of very variable size and clearly visible. Wingspan: 21–28 mm.

Genitalia (♂): (Figs. 7.65A–B). Preanal appendages with an upper subapical tooth and their hooked lower apical angle. Parameres robust, a little shorter than the phallus. Second segment of the inferior appendages with its upper triangular subapical angle and a right-angled notch.

Genitalia (♀): Not available.

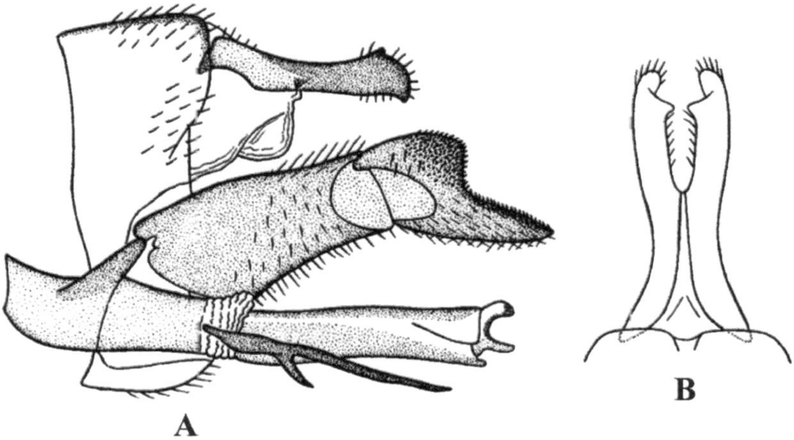

Figure 7.65 *Rhyacophila jayadurga* (♂): (A) Male genitalia, lateral view; (B) Dorsal view of preanal appendage (from Schmid, 1970).

Holotype: INDIA: Madras, Makurti 1-I-1959.

Holotype depository: ROM, Canada.

Distribution: India (Chennai).

59. ***Rhyacophila kadampa*** **Schmid**

Rhyacophila kadampa, Schmid, 1970: 139

Adult: Above of the body dark brown, with concolorous hairiness. Weeping and light red legs. Forewings dark brown and riddled with golden macules, indistinct, but larger and clearly visible in the anal area: Wingspan 20 mm.

Genitalia (♂): (Figs. 7.66A–B). Apical lobe of IXth segment very strong and not very indented at its extremity. Dorsal appendage of the phallic apparatus slightly incised; parameres thick, but sharply thinned and curved upwards on their apical third and ending in a few obtuse teeth. Second segment of the inferior appendages quite large and sub-rectangular in shape.

Genitalia (♀): Not available.

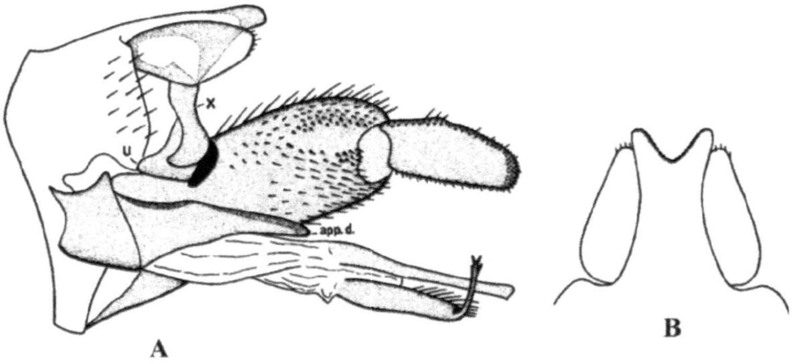

Figure 7.66 *Rhyacophila kadampa* (♂): (A) Male genitalia, lateral view; (B) Dorsal view (from Schmid, 1970).

Holotype: INDIA: Manipur; Sirohi Kashong, 6-7-VI-1960.

Holotype depository: ROM, Canada.

Distribution: India (Manipur).

60. ***Rhyacophila kadaphes*** **Schmid**

Rhyacophila kadaphes, Schmid, 1959: 235–236

Adult: Head, thorax, abdomen and legs brown; tibia of foreleg and midleg with distinct markings; maxillary and labial palp pale brown. Length of antenna 9.2 mm and forewing 12.0 mm. In forewing fork 2nd arise slightly earlier than 1st.

Genitalia (♂): (Figs. 7.67A–C). In dorsal view, dorsal apical lobe of segment IX broader at apex and gradually goes on decreasing up to base. In lateral view, tergum of segment IX longer than its sternum, dorsal apical lobe of segment IX sub-truncate apically. Preanal appendage broader at base and rounded apically. Anal sclerite bilobed. Phallicata expanded apically and ventral lobe of phallicata irregularly shaped apically. Paramere broader at middle. Basal segment of inferior appendage longer than the last segment. Inferior lobe of terminal segment of inferior appendage slightly longer than superior lobe. From superior side basal segment of inferior appendage with slightly concavity at middle. In ventral view inferior lobe of terminal segment of inferior appendage rounded.

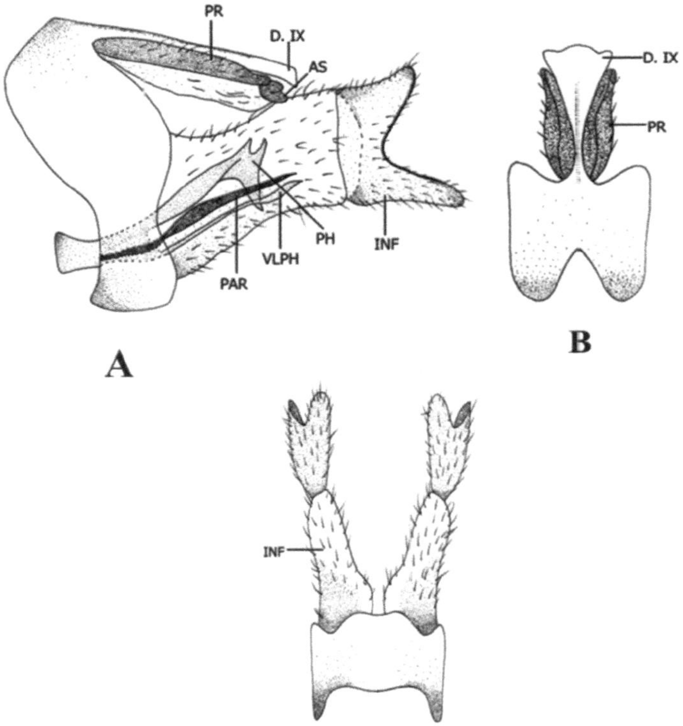

Figure 7.67 *Rhyacophila kadaphes* (♂): (A) Lateral view of male genitalia; (B) Dorsal view; (C) Ventral view.

Genitalia (♀): (Figs. 7.68A–E). Laterally, VIIIth segment simple in shape and dorsal apodeme extends anteriorly until Vth segment. Basal apodeme arise from VIth segment and extend posteriorly up to segment VII and slightly bifurcated at apex. Tergum of VIIIth segment simple in outline and without any indentation along the posterior border, sternum with a shallow indentation.

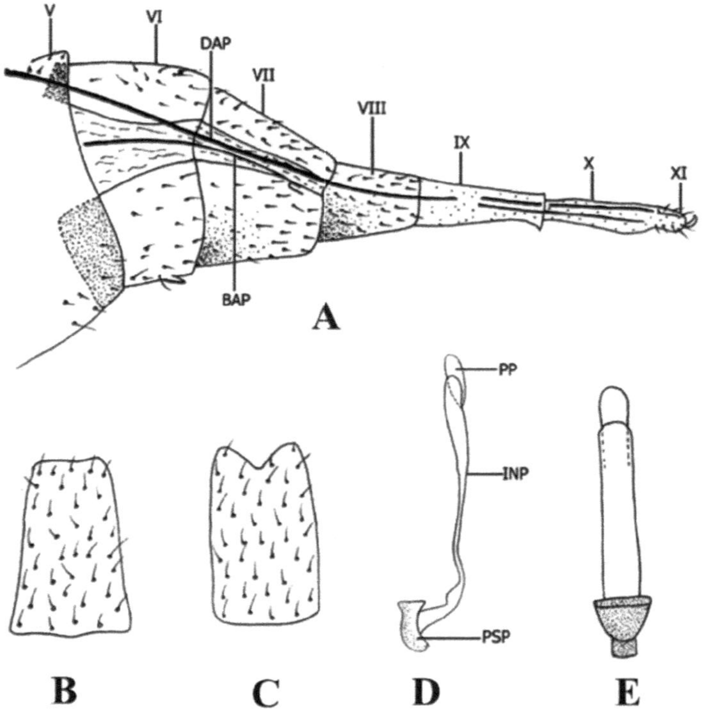

Figure 7.68 *Rhyacophila kadaphes* (♀): (A) Lateral view of female genitalia; (B) Dorsal view of segment VIII; (C) Ventral view of segment VIII; (D) Lateral view of vaginal apparatus; (E) Ventral view of vaginal apparatus.

Holotype: PAKISTAN: Himalaya; Besal, Dunga Nar; 8-27-VII-1953.

Holotype depository: Zoological Museum of Lausanne, Switzerland.

Additional material examined: INDIA: Jammu & Kashmir: Hunder, 3200 m, 05-VIII-2008; Sonmarg, 2700 m, 11-VIII-2008;

Sonmarg, 2700 m, 05-VIII-2009; Pahalgam, 2100 m, 12-VIII-2010; Sonmarg, 2700 m, 26-VI-2011; Himachal Pradesh: Marhi, 2700 m, 03-VIII-2008; Losar, 3200 m, 22-VIII-2008; Kaza, 3000 m, 24-VIII-2008; Solang Valley, 2800 m, 05-VII-2009; Uttarakhand: Janki Chatti, 2600 m, 25-IX-2008; Shyna Chatti, 2000 m, 27-IX-2008, Mana, 3200 m, 21-VI-2012. (Coll. Parey & Pandhar).

Distribution: India/ Pakistan (Uttarakhand, Jammu and Kashmir, Himachal Pradesh).

61. *Rhyacophila kagyupa* Schmid

Rhyacophila kagyupa, Schmid, 1970: 197

Adult: Upper body light reddish-brown with golden hair-iness. Pleura and legs reddish yellow. Forewings brown and riddled with indistinct golden blotches. Wingspan: 17–23 mm.

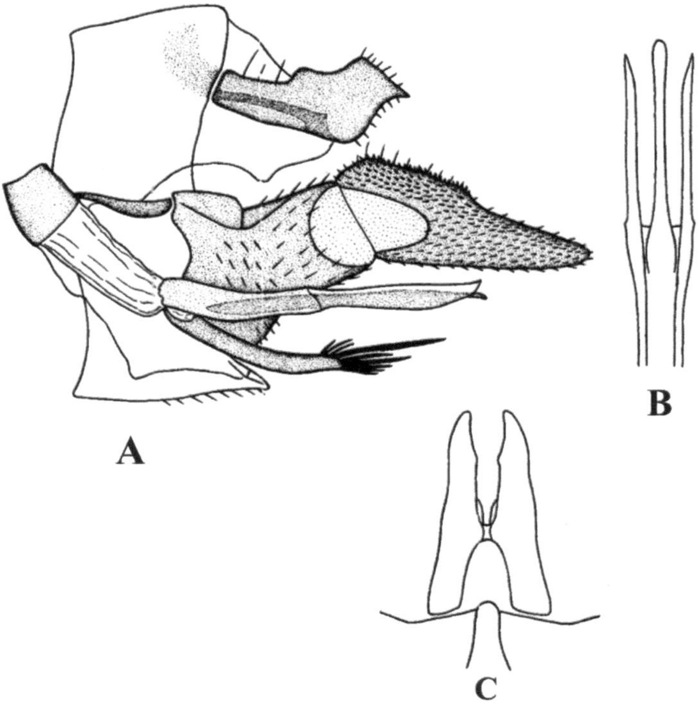

Figure 7.69 *Rhyacophila kagyupa* (♂): (A) Lateral view of male genitalia; (B) Dorsal view phallus; (C) Preanal appendage, dorsal view (from Schmid, 1970).

Genitalia (♂): (Figs. 7.69A–C). Broad, convoluted preanal appendages. phallus slender and terminated in three very long, slender branches situated on the same plane; parameres short, stout and armed with a bunch of long apical spines. Second segment of the inferior appendage very elongated.

Genitalia (♀): Not available.

Holotype: PAKISTAN: Assam; NEFA, Kameng Frontier Division, Rupa, 5500–8800 ft., 11 VI-1961.

Holotype depository: ROM, Canada.

Distribution: India (Arunachal Pradesh).

62. *Rhyacophila kando kando* Schmid

Rhyacophila kando, Schmid, 1970: 206–207

Subspecies: *Rhyacophila kando rengma,* Schmid, 1970: 207

Adult, *Rhyacophila kando*: Dorsal side of body reddish-brown, with golden hairs. Pleura and legs reddish yellow. Very small claws; anterior and middle very large. Forewings of a fairly dark chestnut brown, finely and very indistinctly spotted; along the post-costal edge, there is a clearly visible light beige band with a slightly sinuous edge.

Genitalia, *Rhyacophila kando* (♂): (Figs. 7.70A–B). apical dorsal lobe of IXth segment appearing very thin, seen laterally and in a pointed triangle, seen dorsally. Anal sclerites very small, but very prominent and always visible in lateral view. Aedeagus (phallus) very long, swollen and pleated longitudinally in its median part; parameres slender, but gradually widening to the apex which is obliquely truncated and finely toothed.

Holotype: INDIA: Assam, NEFA, Kameng Frontier Division, Phutang, 1-4-X-1961.

Holotype depository: ROM, Canada.

Additional material examined: INDIA: Arunachal Pradesh; Lumla, 2300 m, 15-V-2011; Uttarakhand; Mornala, 2200 m, 11-VI-2008; Pothibasa, 2200 m, 15-VI-2010; Badrinath, 3400 m, 22-VI-2011; Pingla Pani, 1800 m, 11-VI-2011. (Coll. Parey & Pandhar).

Distribution: Bhutan, Nepal, India (Arunachal Pradesh, Uttarakhand).

Genitalia, *Rhyacophila kando rengma* (♂): (Figs. 7.70C–F). Apical dorsal lobe of segment IX appearing almost as high as it is long, seen dorsally and clearly truncated at its end, seen from above. Tiny anal sclerites entirely hidden within the Xth segment.

Genitalia (♀): Not available.

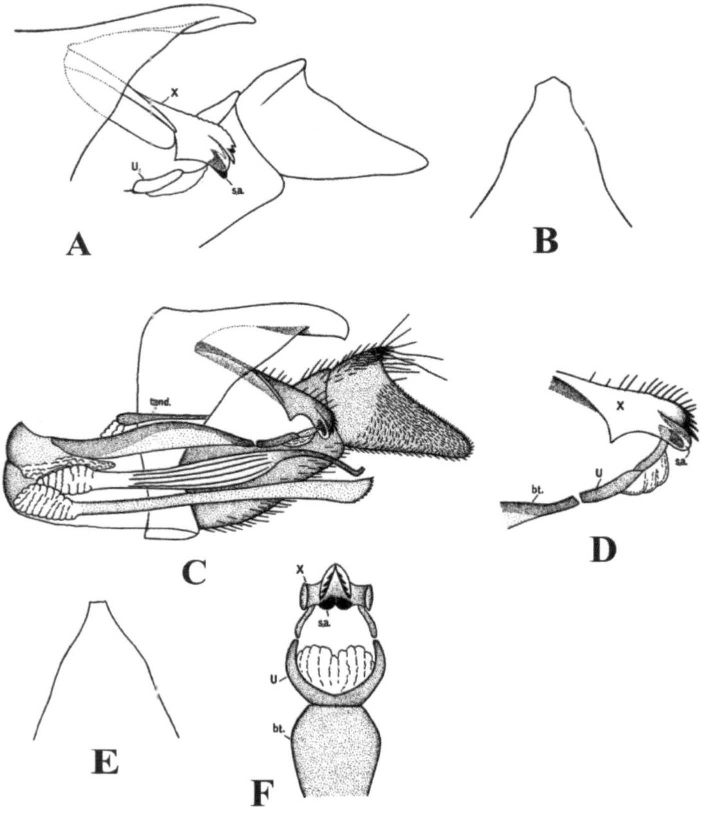

Figure 7.70 (A) *Rhyacophila kando* (♂): Male genitalia, lateral view; (B) Apical dorsal lobe of IXth segment, dorsal view; (C) *Rhyacophila kando rengma* (♂), lateral view of male genitalia; (D) Lateral view of IXth segment and tergal assembly; (E) Ventral view of IXth segment and tergal assembly; (F) Apical dorsal lobe of IXth segment, dorsal view (from Schmid, 1970).

Holotype: INDIA: Manipur, Hkayam Boum, 7300–8000 ft., 20-23-VI-1960.

Holotype depository: ROM, Canada.

Distribution: India (Manipur).

63. *Rhyacophila kangjongpa* Schmid

Rhyacophila kangjongpa, Schmid, 1970: 163

Adult: Dorsal side of the body black, with dark brown hairs. Pleura and legs brown. Uniformly black wings. Wingspan: 13–14 mm.

Genitalia (♂): (Figs. 7.71A–B). When seen laterally, dorsal part long, narrow and appearing slightly sinuous; seen from above, it appears indented in a triangle at its end and ends with two small teeth directed downwards. Anal sclerite short, wide and barely incised at its apical edge. Second segment of inferior appendages much shorter than high and with its upper apical angle a little elongated.

Genitalia (♀): Not available.

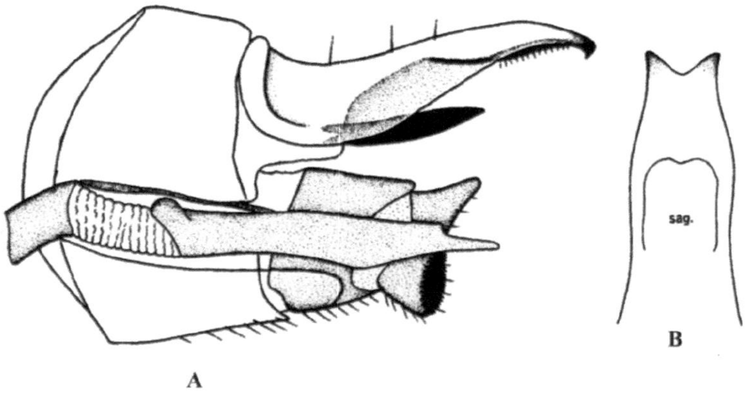

Figure 7.71 *Rhyacophila kangjongpa* (♂): (A) Lateral view of male genitalia; (B) Dorsal view (from Schmid, 1970).

Holotype: INDIA: Assam, NEFA, Kameng Frontier Division, Kujjalong 28-30-VI-1960.

Holotype depository: ROM, Canada.

Distribution: India (Assam, Arunachal Pradesh).

64. *Rhyacophila kanichka* Schmid

Rhyacophila kanichka, Schmid, 1959: 233–234

Adult: Head and thorax brown; antenna, maxillary palp and labial palp, abdomen, legs and wings pale brown. Length of antenna 8.73 mm and forewing 13.5 mm. In forewing forks 1st and 2nd arise at the same level.

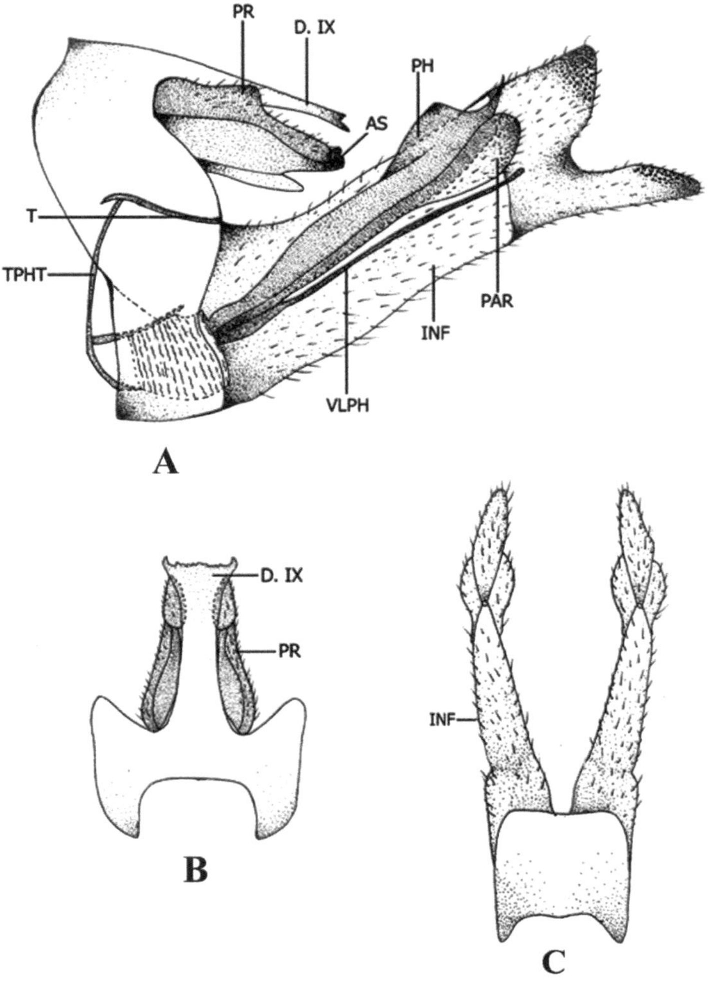

Figure 7.72 *Rhyacophila kanichka* (♂): (A) Lateral view of male genitalia; (B) Dorsal view; (C) Ventral view (from Schmid, 1970).

Genitalia (♂): (Figs. 7.72A–C). In dorsal view, apex of dorsal apical lobe of segment IX with serrated margin. Segment IX deeply excised above; the center of excision produced in a long process. Inferior appendage long and breadth of basal segment almost same, terminal segment with deep rounded indentation, superior lobe quadrate, wider and shorter than inferior lobe. Preanal appendage hinged at the base of dorsal apical lobe of segment IX and apically rounded. Anal sclerite bilobed. Phallicata long, ventral lobe thin and as long as phallicata and apically rounded. Paramere long spinifere broad at apex. In ventral view basal segment of inferior appendage broad at base.

Genitalia (♀): Not available.

Holotype and Allotype: PAKISTAN: Lilam, 9000–11000 ft., 24-V-1954, Saif-ul-Maluk Sar, 1-3-VII-1953, Besal, 8-9-VII-1953.

Holotype depository: Zoological Museum of Lausanne, Switzerland.

Additional material examined: INDIA: Uttarakhand: Gobinddham, 3100 m, 22-VI-2008; Gobinddham, 3100 m, 18-VI-2012. (Coll. Parey & Pandhar).

Distribution: India/Pakistan (Uttarakhand).

65. *Rhyacophila karpa* Schmid

Rhyacophila karpa, Schmid, 1970: 204

Adult: Dorsal side of the body dark reddish-brown, with concolorous hairs. Pleura and femurs reddish yellow; tibiae and tarsi brown. Forewings dark brown, with a few clear spots, clearly visible, but with not very clear contours. Wingspan: 13.5–16 mm.

Genitalia (♂): (Figs. 7.73A–B). Apical dorsal part of segment IX with a dense brush of thickened hairs. Dorsal part of the Xth segment very long, with an upper preapical bulge and its end curved into a hook. Second segment of the inferior appendages long triangular and with its slightly concave upper edge.

Genitalia (♀): Not available.

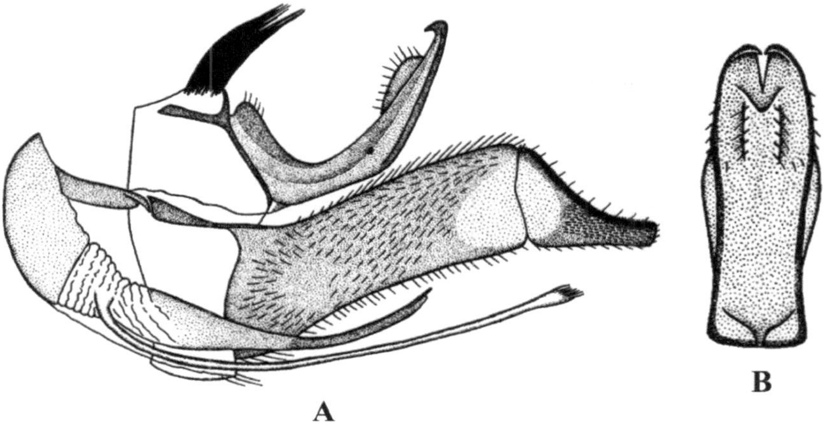

Figure 7.73 *Rhyacophila karpa* (♂): (A) Lateral view of male genitalia; (B) Dorsal view (from Schmid, 1970).

Holotype and Allotype: INDIA: Assam; United Jaintia and Khasi Hills, Tkentalong, 29-31-XII-1959.

Holotype depository: ROM, Canada.

Distribution: India (Meghalaya).

66. *Rhyacophila kashmirensis* Kaur and Saini

Rhyacophila kashmirensis, Kaur and Saini, 2013

Adult: Body dark brown, wings and legs pale brown. Length of antenna 8.2 mm and length of forewing 9.5 mm. In forewing fork 2nd arise slightly earlier than 1st.

Genitalia (♂): (Figs. 7.74A–E). In dorsal view, dorsal apical lobe of segment IX with U shape indentation; in lateral view, anal sclerite extending beyond dorsal apical lobe of segment IX. Phallicata bifid and bunch of setae present on apex of paramere. Dorsal appendage of phallicata rounded apically and with minute spinules. Basal segment of inferior appendage broader at base. In lateral view, terminal segment of inferior appendage truncates but rounded in ventral view. In dorsal view, dorsal appendage of phallicata indented subapically.

Genitalia (♀): Not available.

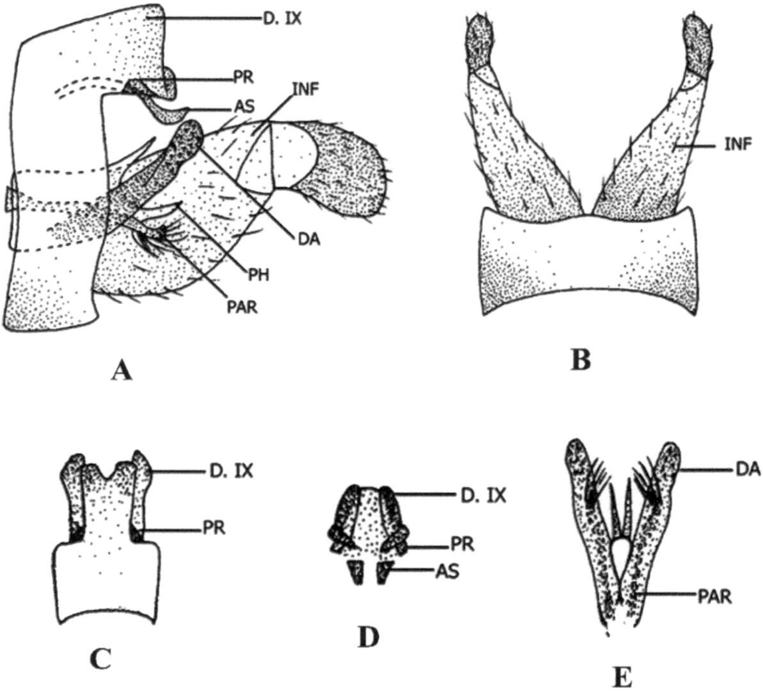

Figure 7.74 *Rhyacophila kashmirensis* (♂): (A) Lateral view of male genitalia; (B) Ventral view; (C) Dorsal view; (D) Caudal view; (E) Dorsal view of dorsal appendage of Phallicata.

Holotype: INDIA: Jammu & Kashmir, Gulmarg, 2900 m, 04-VII-2012. (Coll. Parey & Pandhar).

Holotype depository: MDZPUP, India.

Distribution: India (Jammu and Kashmir).

67. ***Rhyacophila kashongpa*** **Schmid**

Rhyacophila kashongpa, Schmid, 1970: 148–149

Adult: Top of the head reddish yellow, with light brown hairs. Upper thorax brown, with a silver midline. Pleura and legs yellow. Relatively broad wings: the front ones are dark brown, intensely and indistinctly mottled with gold. Wingspan: 20–21 mm.

Genitalia (♂): (Figs. 7.75A–D). Xth segment small, sub-horizontal and terminated by two large, rounded lobes. Large and subcircular anal sclerites. Second segment of the inferior appendages in an oblique triangle downwards and with a very slight apical indentation. Phallic apparatus very long and slender; phallus composed of two thin branches of unequal length; ventral plate very long and terminated by two rounded lobes.

Genitalia (♀): Not available.

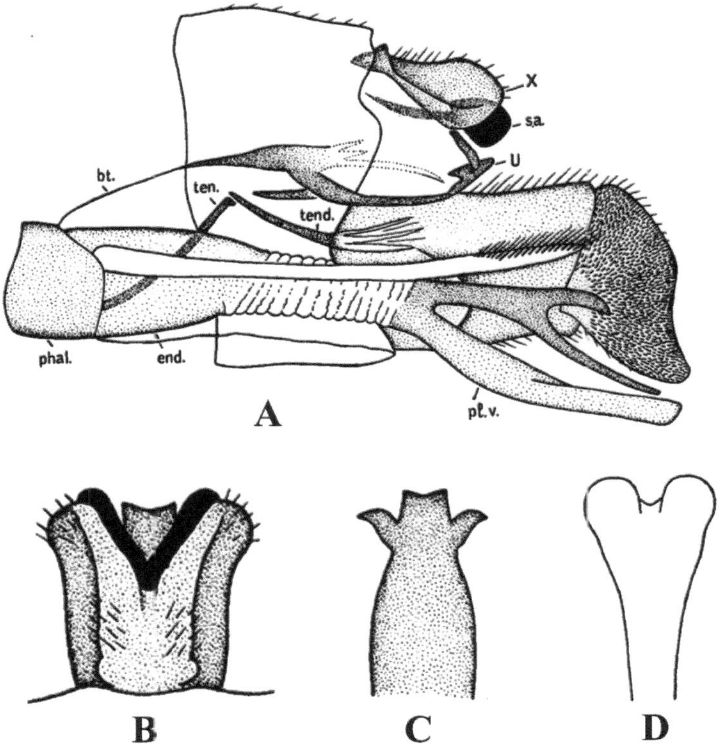

Figure 7.75 *Rhyacophila kashongpa* (♂): (A) Male genitalia, lateral view; (B) Dorsal view of Xth segment; (C) Tergal assembly, ventral view; (D) Ventral plate of phallus (Aedeagus), ventral view (from Schmid, 1970).

Holotype: INDIA: Manipur, Sirohi Kashong, 6800–7500 ft., 6-7-VI-1960.

Holotype depository: ROM, Canada.

Distribution: India (Manipur).

68. ***Rhyacophila kawachenpa*** **Schmid**

Rhyacophila kawachenpa, Schmid, 1970: 163–164

Adult: Dorsal side of the body dark brown, with golden hairs. Pleura and legs light reddish yellow. Wings not very large in relation to the body; the front ones are brown, with a few large, inconspicuous golden macules. Wingspan: 15–24 mm.

Genitalia (♂): (Figs. 7.76A–B). Dorsal piece very long, narrow, truncated at its extremity and without lower teeth. Anal scleritis also long, narrow and distinctly bilobed. Sagittal appendage short and very broad. Second segment of the inferior appendages as long as it is high and with its distinctly prominent upper apical angle.

Genitalia (♀): Not available.

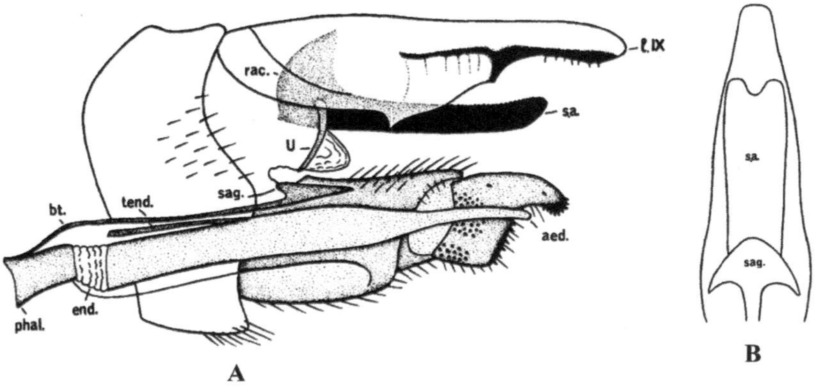

Figure 7.76 *Rhyacophila kawachenpa* (♂): (A) Lateral view of male genitalia; (B) Dorsal view (from Schmid, 1970).

Holotype: INDIA: Assam, NEFA, Kameng Frontier Division, Domkho 11-V-1961.

Holotype depository: ROM, Canada.

Distribution: India (Arunachal Pradesh).

69. ***Rhyacophila kedara*** **Schmid**

Rhyacophila kedara, Schmid, 1970: 170

Adult: Entire body brown except hindwing, maxillary and labial palpi which are pale brown; antenna with alternating dark and light brown bands; tibia of foreleg and midleg with distinct markings. Length of antenna 11.64 mm and forewing 13.5 mm. In forewing forks 1st and 2nd arise at the same level.

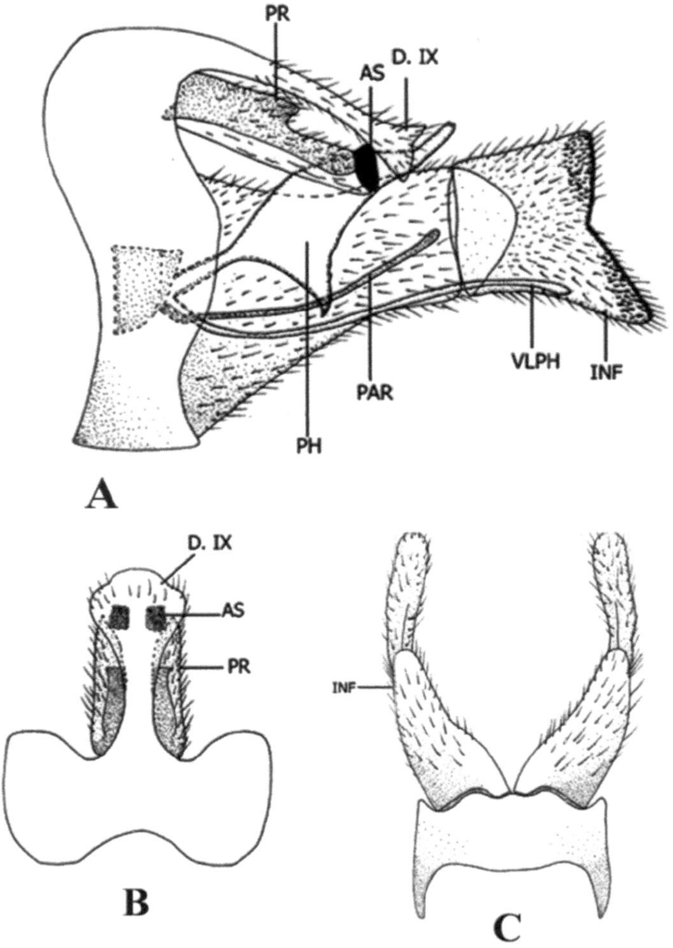

Figure 7.77 *Rhyacophila kedara* (♂): (A) Lateral view of male genitalia; (B) Dorsal view; (C) Ventral view.

Genitalia (♂): (Figs. 7.77A–C). In dorsal view, dorsal apical lobe of segment IX narrow at base, gradually widening towards apex. In lateral view, inferior appendage long, basal segment somewhat tapered from base to apex; terminal segment with sharp excision almost at right angles, inferior lobe long. Preanal appendage shorter than dorsal apical lobe of segment IX, broader at base and narrow apically, dorsally long in-folded. Anal sclerite thickly sclerotized and small, dorsally hidden beneath dorsal apical lobe of segment IX. Phallicata as long as basal segment of inferior appendage, bifid apically with upper lobe long and rounded and lower lobe short with pointed apex, ventral lobe of phallicata slender, little longer than phallicata; paramere with rounded apex. In ventral view terminal segment of inferior appendage rounded apically.

Genitalia (♀): Not available.

Holotype: INDIA: Uttarakhand; Pauri Garhwal, Kedarnath, 6900–13000 ft., 13-15-V-1958.

Holotype depository: ROM, Canada.

Additional material examined: INDIA: Jammu & Kashmir: Gulmarg, 2900 m, 17-IX-2012; Uttarakhand: Chopta, 2700 m, 13-VI-2009. (Coll. Parey & Pandhar).

Distribution: India (Arunachal Pradesh).

70. ***Rhyacophila khamakhya*** **Schmid**

Rhyacophila khamakhya, Schmid, 1970: 197

Adult: Dorsal side reddish-brown, covered with abundant burnt gold hair. Pleura and legs reddish yellow. Forewings reddish gold, with few, irregular brown macules, but especially visible along the apical and post-costal edges. There are two brown transverse bands, one over the anastomosis and one in the middle of the apical cells. Wingspan: 17–20 mm.

Genitalia (♂): (Figs. 7.78A–B). Strong preanal appendages, quite high and with a clear indentation at the upper apical angle. Parameres strong and with two adjoining teeth. Upper lobe of the second segment of inferior appendages somewhat variable, but always angular, slightly curved outwards and sometimes pointed.

Genitalia (♀): Not available.

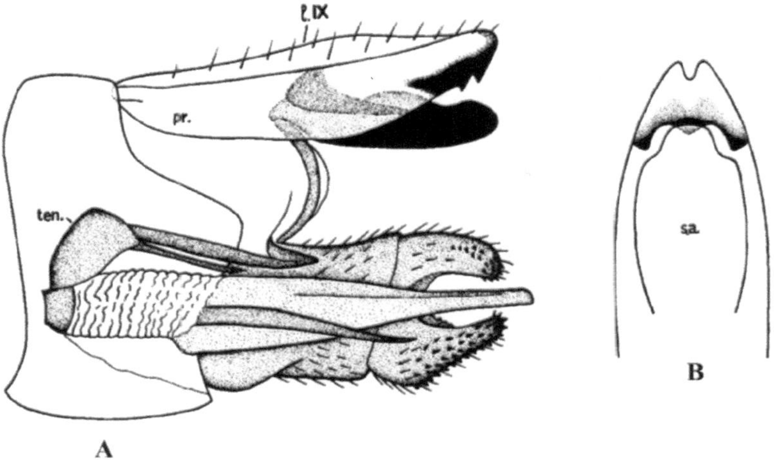

Figure 7.78 *Rhyacophila khamakhya* (♂): (A) Lateral view of male genitalia; (B) Dorsal view (from Schmid, 1970).

Holotype: INDIA: Madras, Kodaikanal, 5000–8000 ft., 6-XII-1961.

Holotype depository: ROM, Canada.

Distribution: India (Arunachal Pradesh).

71. ***Rhyacophila khasiorum* Schmid**

Rhyacophila khasiorum, Schmid, 1970: 187–188

Adult: Entire body brown; antenna with alternating light and dark brown bands; tibia and tarsus of foreleg, mid leg entirely and only tarsus of hind leg dark brown. Length of antenna 4.3 mm and forewing 6.79 mm. In forewing fork 2nd arise earlier than 1st.

Genitalia (♂): (Figs. 7.79A–B). In dorsal view, dorsal apical lobe of segment IX rounded and preanal appendage pointed apically. Anal sclerite truncate in outline. In lateral view, dorsal apical lobe of segment IX smaller than preanal appendage. Dorsal appendage of phallic apparatus rounded in outline. Phallicata bilobed and lower part longer than the upper part. Phallicata with small ventral lobe. Paramere pointed and between the two parts of phallicata. Terminal segment of inferior appendage smaller than the basal segment of inferior appendage. In ventral view, bases of basal segment of inferior appendage protruding and broader than the rest part of basal segment of inferior appendage.

Genitalia (♀): Not available.

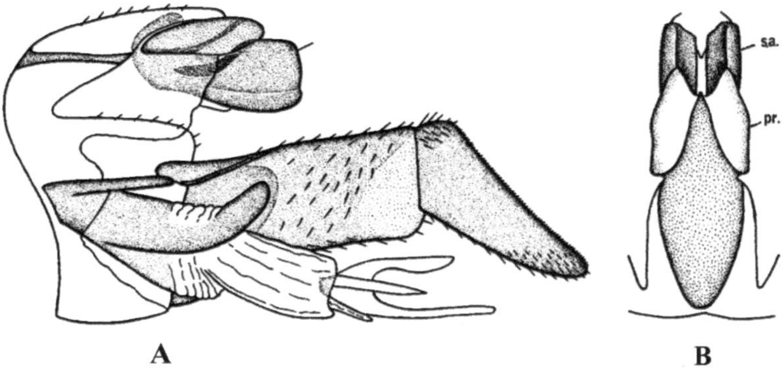

Figure 7.79 *Rhyacophila khasiorum* (♂): (A) Lateral view of male genitalia; (B) Dorsal view (from Schmid, 1970).

Holotype: INDIA: Assam; United Jaintia and Khasia Hills, Rumkheng, 1500–9000 ft., 20-23-III-1959.

Holotype depository: ROM, Canada.

Additional material examined: INDIA: Himachal Pradesh; Satrundi, 2700 m, 15-VI-2011; Uttarakhand; Auli, 2900 m, 24-VI-2012. (Coll. Parey & Pandhar).

Distribution: India (Arunachal Pradesh).

72. *Rhyacophila khimbarpa* Schmid

Rhyacophila khimbarpa, Schmid, 1970: 188

Adult: Dorsal side of the body reddish-brown, with concolorous hairiness. Pleura and legs light reddish yellow, with the apex of the median and posterior hips very light. Wings uniformly black-brown. Wingspan: 11.5–20 mm.

Genitalia (♂): (Figs. 7.80A–B). Preanal appendages oval and slightly indented at their extremity. Massive anal sclerites, particularly high and short. Dorsal appendages of the phallic apparatus broad and rounded at their extremity; phallic apparatus small, but with all its parts equally developed.

Second segment of the inferior appendages sub-quadrangular, wider at its extremity than at its base and with its apical edge regularly convex.

Genitalia (♀): Not available.

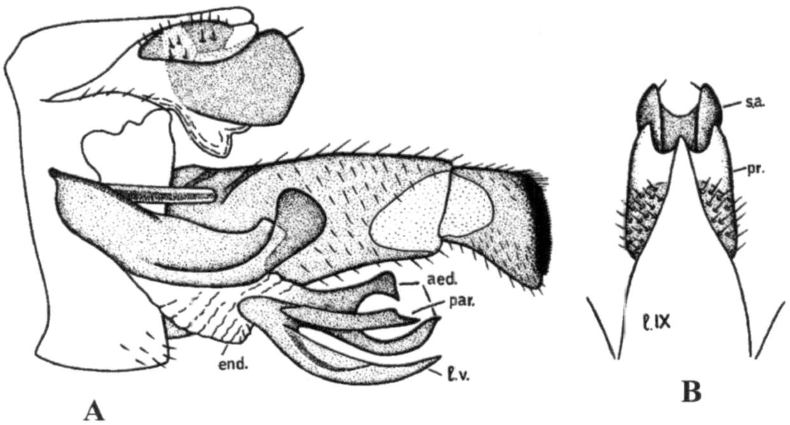

Figure 7.80 *Rhyacophila khimbarpa* (♂): (A) Lateral view of male genitalia; (B) Dorsal view (from Schmid, 1970).

Holotype: INDIA: Assam, NEFA, Kameng Frontier Division, Jhum La, 7200–8800 ft., 13-V-1961.

Holotype depository: ROM, Canada.

Distribution: India (Assam, Arunachal Pradesh).

73. *Rhyacophila khiympa* Schmid

Rhyacophila khiympa, Schmid, 1970: 170–171

Adult: Dorsal side of the body reddish-brown, with concolorous hairs. Yellow-brown pleura and legs. Anterior and median tibiae and tarsi brown, ringed with light. Wings narrow and dark in color. The front ones are fairly lightly mottled, except in the thyroid area, where macules are visible, light and dark, larger than the others.

Genitalia (♂): (Figs. 7.81A–B). Apical dorsal lobe of segment IX small. Preanal appendages very high and notched at their extremity. Aedeagus (phallus) slender in shape, with its strong ventral lobes ending in ovals; remarkably long parameres and slender, spiniform, and recurved. Second segment of inferior appendages terminated in two sub-equal lobes.

Genitalia (♀): Not available.

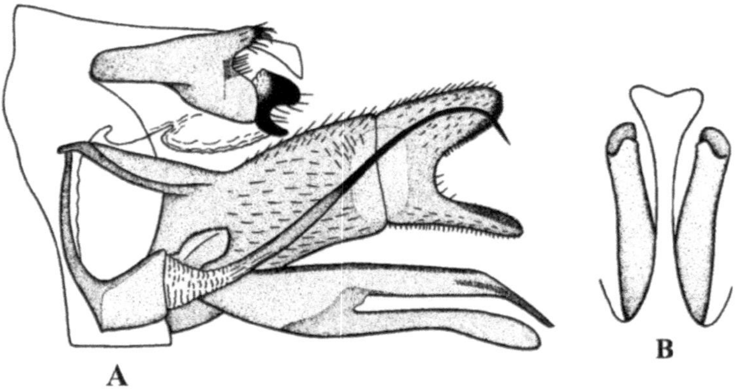

Figure 7.81 *Rhyacophila khiympa* (♂): (A) Lateral view of male genitalia; (B) Dorsal view (from Schmid, 1970).

Holotype: INDIA: Sikkim; Bakkim, 6900–1400 ft., 12-IV-1959.

Holotype depository: ROM, Canada.

Distribution: China, Bhutan, India (Sikkim).

74. *Rhyacophila kubra* Schmid

Rhyacophila kubra, Schmid, 1970: 178

Adult: Top of the body black with concolorous hairs. Dark brown pleura and legs. Uniformly black wings. Bright orange abdomen. Wingspan: 15–17 mm.

Genitalia (♂): (Fig. 7.82). Apical dorsal lobe of the IXth segment very long, strongly convex towards the top and slightly concave at its end. Preanal appendages very elongated ovals and entirely joined to the preceding one. Anal sclerites apparently disappeared. Phallic apparatus reduced to the phallus which is composed of two short branches: a dorsal one, broadly oval and a ventral one, cylindrical, and to the parameres, long sickle spines, slightly spiniferous. Second segment of the inferior appendages small, flared and not indented.

Genitalia (♀): Not available.

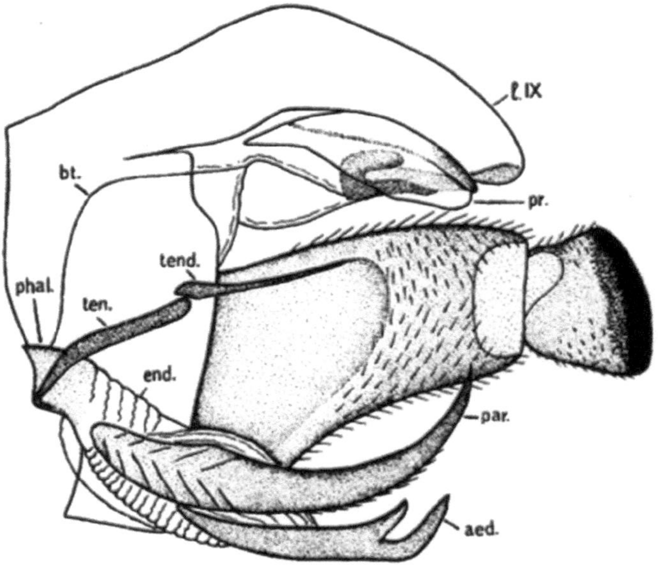

Figure 7.82 *Rhyacophila kubra* (♂): Lateral view of male genitalia (from Schmid, 1970).

Holotype: INDIA: Sikkim; Chateng, 6200–8700 ft., 12-VI-1959.

Holotype depository: ROM, Canada.

Distribution: Nepal, Bhutan, India (Sikkim).

75. *Rhyacophila kunma* Schmid

Rhyacophila kunma, Schmid, 1970: 154–155

Adult: Dorsal side of the body black-brown. Dark brown pleura and legs. Wingspan: 19 mm.

Genitalia (♂): (Fig. 7.83). Horizontal part of Xth segment not very long; vertical part armed with two top pairs, very short and triangular points and a lower point of the same size and the same shape, but odd. Inferior appendages not very long and with non-straight edges. Second segment of inferior about twice as long as high and obliquely truncated at its end.

Genitalia (♀): Not available.

Holotype: INDIA: Assam; NEFA, Kameng Frontier Division, Rahung, 7000 ft., 25-IV-1961.

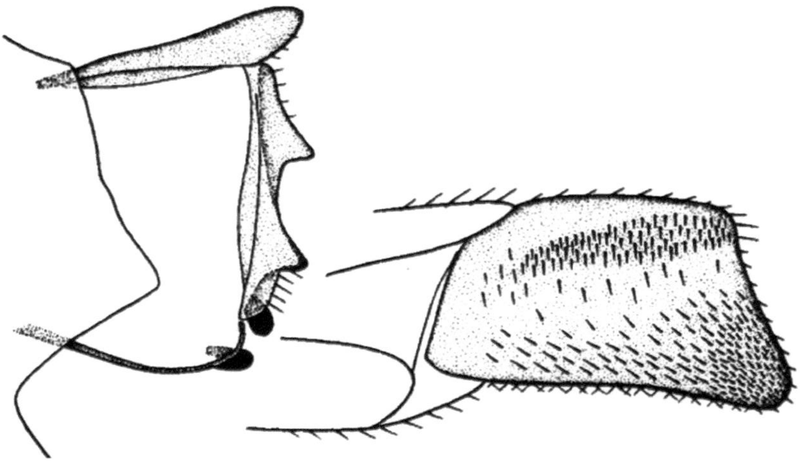

Figure 7.83 *Rhyacophila kunma* (♂): Lateral view of male genitalia (from Schmid, 1970).

Holotype depository: ROM, Canada.

Distribution: India (Assam, Arunachal Pradesh).

76. *Rhyacophila kusang* Schmid

Rhyacophila kusang, Schmid, 1970: 145

Adult: Dorsal side of the body dark reddish-brown, with concolorous hairs. Pleuron and legs reddish yellow. Forewings very dark golden tinder, riddled with golden and silvery macules, inconspicuous. Wingspan: 15 mm.

Genitalia (♂): (Figs. 7.84A–D). Xth segment small and armed with two caudal spines arranged in a pointed arch. Second segment of inferior appendages small, simple and slightly indented at its extremity. Aedeagus (phallus) very thick over its entire length, with its dorsal branch reduced; very short ventral plate arms.

Genitalia (♀): Not available.

Holotype: INDIA: Assam; United Jantia and Khasia Hills, Shnongpdeng, 500–2500 ft., 6-XII-1959.

Holotype depository: ROM, Canada.

Distribution: India (Meghalaya).

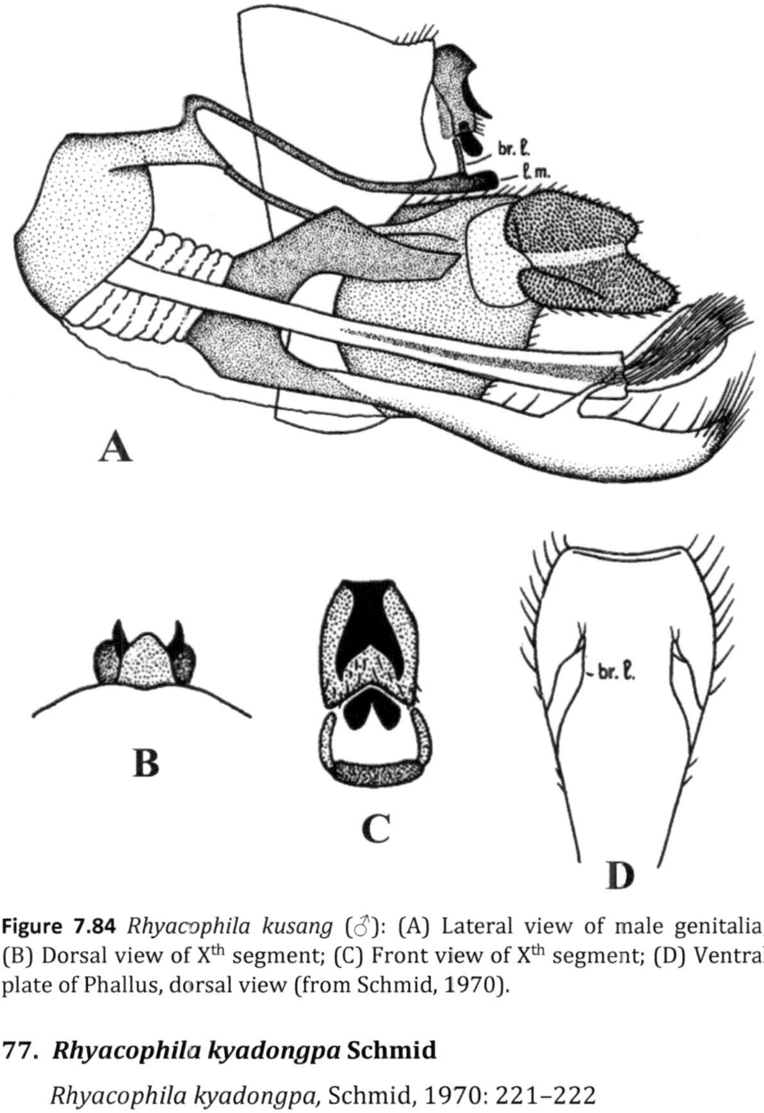

Figure 7.84 *Rhyacophila kusang* (♂): (A) Lateral view of male genitalia; (B) Dorsal view of Xth segment; (C) Front view of Xth segment; (D) Ventral plate of Phallus, dorsal view (from Schmid, 1970).

77. *Rhyacophila kyadongpa* Schmid

Rhyacophila kyadongpa, Schmid, 1970: 221–222

Adult: Dorsal side of the body reddish-brown. Pleuron and legs reddish yellow. Wings long, narrow and not very large. The forewings are golden yellow, riddled with brown macules, larger at the base of the wing and in the post-costal area. Wingspan: 21–25 mm

Genitalia (♂): (Figs. 7.85A–C). Extremely remarkable for the slenderness and sure elegance of the shape of the appendages. The apical dorsal lobe of the IXth segment is composed of three branches: a median shortly bifid and two laterals, very long, slender and gracefully arched. Inferior appendages not very elongated, but very narrow. Phallus small and simple; missing parameres.

Genitalia (♀): Not available.

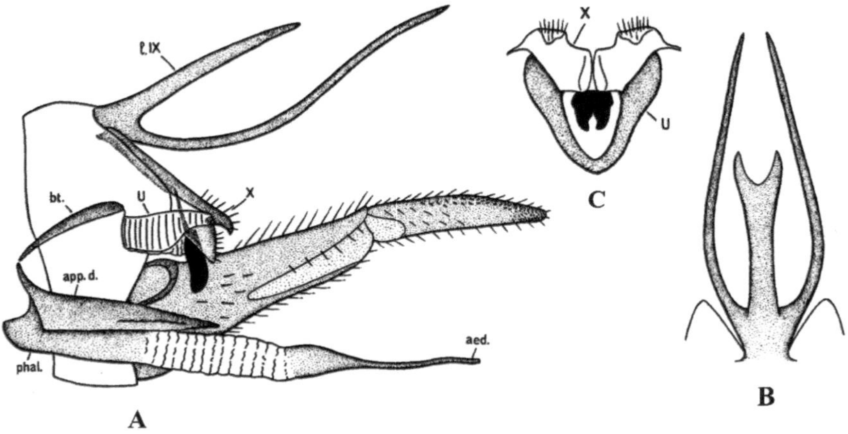

Figure 7.85 *Rhyacophila kyadongpa* (♂): (A) Lateral view of male genitalia; (B) Dorsal apical lobe of IXth segment; (C) Front view of Xth segment (from Schmid, 1970).

Holotype: INDIA: Sikkim; Lachung, 2-13-VII-1959.

Holotype depository: ROM, Canada.

Distribution: India (Sikkim).

78. *Rhyacophila kyimdongpa* Schmid

Rhyacophila kyimdongpa, Schmid, 1970: 176–177

Adult: Top of the body brown black, with concolorous hairs. Yellow-brown pleura and legs. Uniformly black wings. Wingspan 12 mm.

Genitalia (♂): (Figs. 7.86A–D). Preanal appendages large, triangular and wide and posteriorly flared. Anal sclerites located below these and in small ovals. Phallic apparatus very small; ventral plate consisting of two branches, each curving laterally outward at its end. Second segment of inferior appendages large, flared and not indented.

Genitalia (♀): Not available.

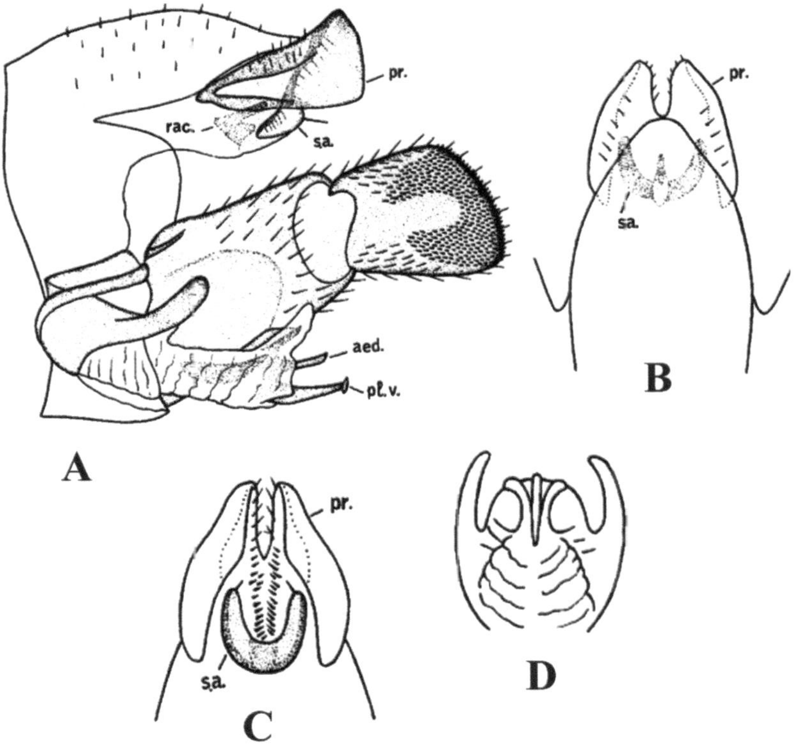

Figure 7.86 *Rhyacophila kyimdongpa* (♂): (A) Lateral view of male genitalia; (B) Dorsal view; (C) Front view of Xth segment; (D) Dorsal view of phallic apparatus (from Schmid, 1970).

Holotype: INDIA: Assam (Now Meghalaya); United Jaintia and Khasia Hills, Serrarim, 5500 ft., 6-7-X-1960.

Holotype depository: ROM, Canada.

Distribution: Thailand, Vietnam, India (Meghalaya).

79. ***Rhyacophila kyungpa*** **Schmid**

Rhyacophila kyungpa, Schmid, 1970: 173

Adult: Top of the body blackish, with concolorous hairs. Dark reddish-brown pleura and legs. Forewings black; both pairs have an area at their tip where the membrane is discolored and band of hair. Abdomen orange. Wingspan: 15–17 mm.

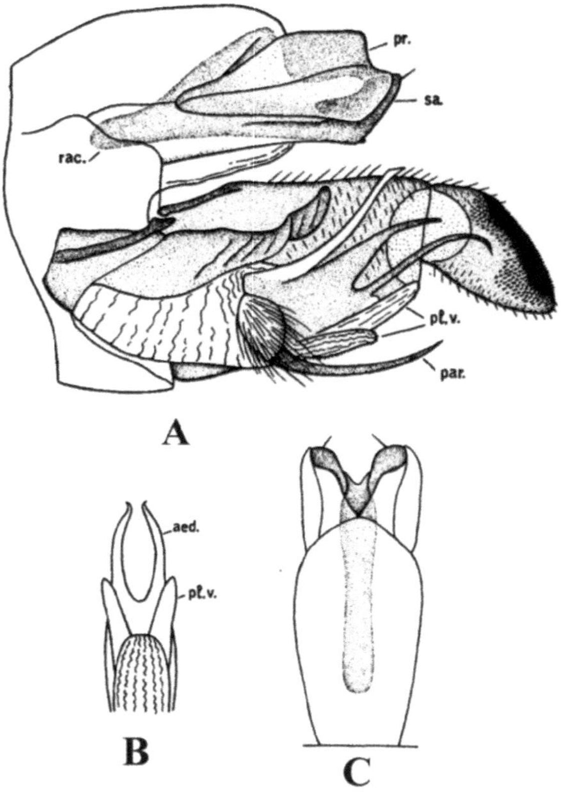

Figure 7.87 *Rhyacophila kyungpa* (♂): (A) Lateral view of male genitalia; (B) Ventral plate of phallus, dorsal view; (C) Dorsal view of IXth and Xth segment (from Schmid, 1970).

Genitalia (♂): (Figs. 7.87A–C). Long preanal appendages. Root of the anal sclerites very stretched. Dorsal appendages of the phallic apparatus in narrow bands, twisted and of irregular relief. Very complex phallus; spiniform parameres, with a membranous, long-hairy basal lobe. Second segment of inferior appendages subovate.

Genitalia (♀): Not available.

Holotype: INDIA: Assam; NEFA, Kameng Frontier Division, Warongpang, 4250–7500 ft., 21-III-1961.

Holotype depository: ROM, Canada.

Distribution: India (Arunachal Pradesh, Assam).

80. *Rhyacophila langdarma* Schmid

Rhyacophila langdarma, Schmid, 1970: 164

Adult: Dorsal side of the body dark reddish-brown, with concolorous hairs. Pleura and legs light reddish yellow. Forewings dark brown, with fairly large, indistinct golden macules. Wingspan: 17–18 mm.

Genitalia (♂): (Figs. 7.88A–B). Fairly high dorsal part, dorsoventrally dilated at its extremity where it bears a circular indentation. Sturdy anal scleritis, acuminate oval. Sagittal appendage very obtuse oval. Second segments of the inferior appendages rhombus shaped, slightly indented at its end.

Genitalia (♀): Not available.

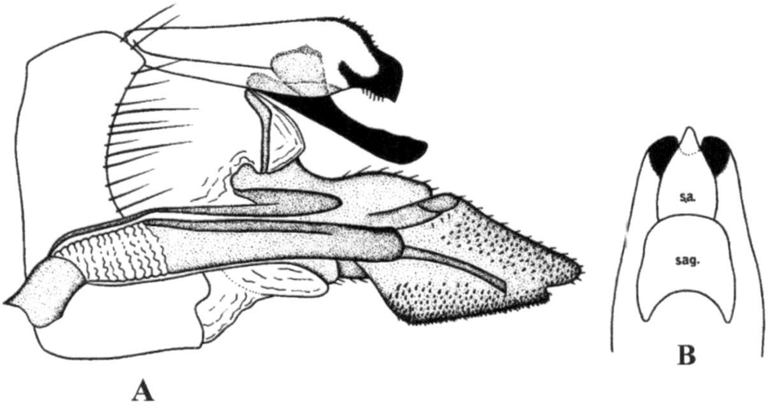

Figure 7.88 *Rhyacophila langdarma* (♂): (A) Lateral view of male genitalia; (B) Dorsal view (from Schmid, 1970).

Holotype: INDIA: Assam; NEFA, Kameng Frontier Division, Dirang Dzong, 6100–7200 ft., 21-22-VII-1961.

Holotype depository: ROM, Canada.

Distribution: India (Arunachal Pradesh, Assam).

81. *Rhyacophila laptsapa* Schmid

Rhyacophila laptsapa, Schmid, 1970: 200–201

Adult: Head, antenna, mesothorax, metathorax, abdomen, wings and legs brown; maxillary palp, labial palp and prothorax pale brown. Length of antenna 6.79 mm and forewing 7.76 mm. In forewing forks 1st and 2nd arise at the same level.

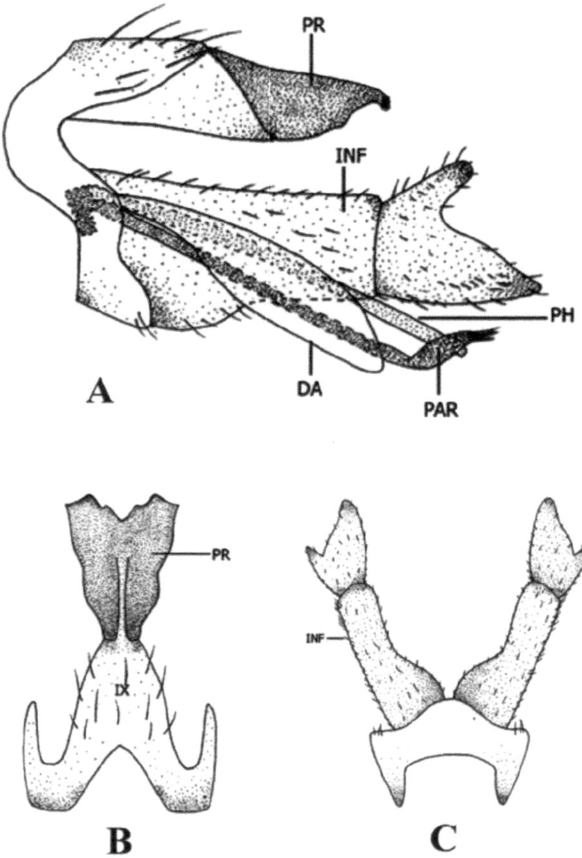

Figure 7.89 *Rhyacophila laptsapa* (♂): (A) Lateral view of male genitalia; (B) Dorsal view; (C) Ventral view.

Genitalia (♂): (Figs. 7.89A–C). In dorsal view preanal appendage indented apically and in lateral view segment IX deeply excised and sinuous, ventral margin slightly oblique. Preanal appendage curved downward apically. Inferior appendage with basal segment long, broad basally and narrow beyond middle; terminal segment short dilating and widely excised apically, inferior lobe larger than superior lobe. Phallicata long, slender and straight; paramere as long as phallicata, apically broader and with tuft of setae. Dorsal appendage of phallic apparatus broader at apex and narrower at base. In ventral view apex of basal segment of inferior appendage half in breadth as compared to the base.

Genitalia (♀): Not available.

Holotype: INDIA: Sikkim; Palam, 100–8700 ft., 25-IV-1959.

Holotype depository: ROM, Canada.

Additional material examined: INDIA: Sikkim; Maltin, 1700 m, 20-V-2011. (Coll. Parey & Pandhar).

Distribution: Nepal, India (Sikkim).

82. *Rhyacophila lepcha* Schmid

Rhyacophila lepcha, Schmid, 1970: 193

Adult: Body and wings dark brown with black pubescence. Length of antenna 7.5 mm and with alternating light and dark brown bands. Fork 1st arise earlier than 2nd whereas in females fork 2nd arise slightly earlier than 1st.

Genitalia (♂): (Figs. 7.90A–C). Preanal appendage in lateral view apically almost rounded and phallicata broader at base and rounded apically. Basal segment of inferior appendage with notch. Paramere sclerotized and with conspicuous spine. Dorsal apical lobe of segment IX equal in length to preanal appendage. In ventral view basal segment of inferior appendage with indent at middle and broader towards proximal end. Terminal segment of inferior appendage with concavity.

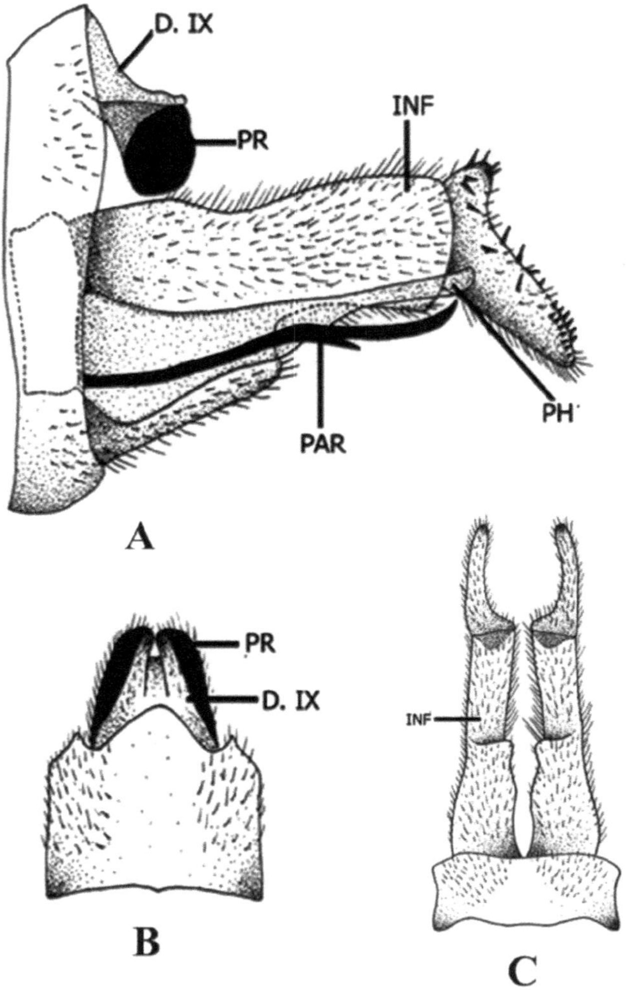

Figure 7.90 *Rhyacophila lepcha* (♂): (A) Lateral view of male genitalia; (B) Dorsal view; (C) Ventral view.

Genitalia (♀): (Figs. 7.91A–E). Tergum VIII slightly concave near posterior margin and segment IX membranous. Dorsal apodeme extending anterad to middle of segment VII and basal apodeme to base of segment VII (Fig. 148). Dorsally segment VIII truncate and sternum trilobed in outline. Vaginal apparatus with sword-shaped posterior process.

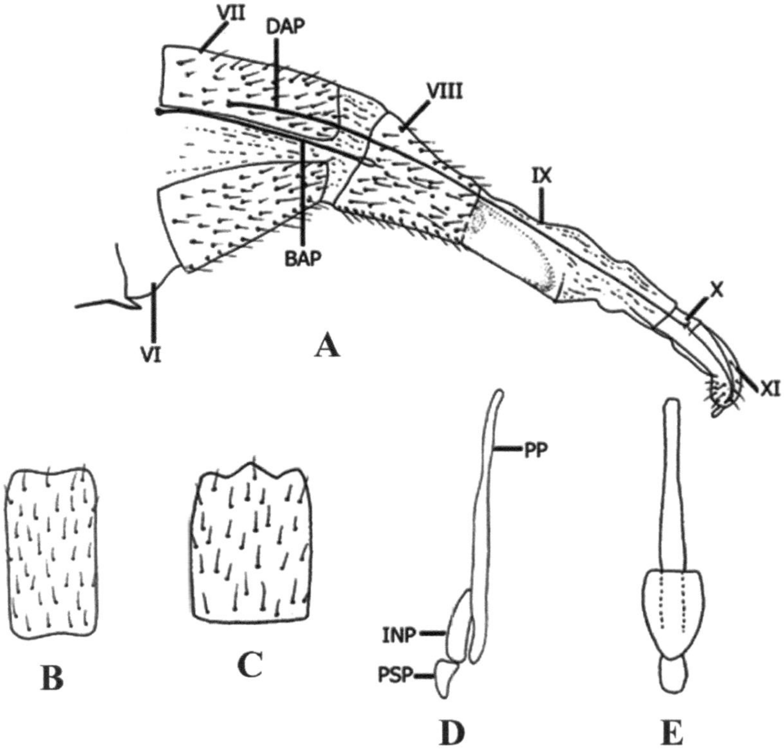

Figure 7.91 *Rhyacophila lepcha* (♀): (A) Lateral view of female genitalia; (B) Dorsal view of segment VIII; (C) Ventral view of segment VIII; (D) Lateral view of vaginal apparatus; (E) Ventral view of vaginal apparatus.

Holotype: INDIA: Sikkim; Zomphuk, 2800–13200 ft., 1-X-1959.

Holotype depository: ROM, Canada.

Additional material examined: INDIA: Sikkim; Lachen, 3200 m, 24-V-2011; Uttarakhand; Pandukeshwar, 3000 m, 20-VI-2012. (Coll. Parey & Pandhar).

Distribution: Bhutan, India (Sikkim, Uttarakhand).

83. *Rhyacophila lhabu* Schmid

Rhyacophila lhabu, Schmid, 1970: 164–165

Adult: Top of the body brown black, with black hairs. Dark brown legs. Forewings with a broad golden spot in the subradial cell, at the level of the anterior tip of the discoid cell. Wingspan: 15–17 mm.

Genitalia (♂): (Figs. 7.92A–C). Dorsal piece appearing straight and acute in lateral view, and long oval in dorsal view; on its underside it bears three obtuse teeth. Subquadrangular anal scleritis and slightly indented at its extremity. Second segment of the inferior appendages longer than high, semi-elliptical and very chitinous at the extremity.

Genitalia (♀): Not available.

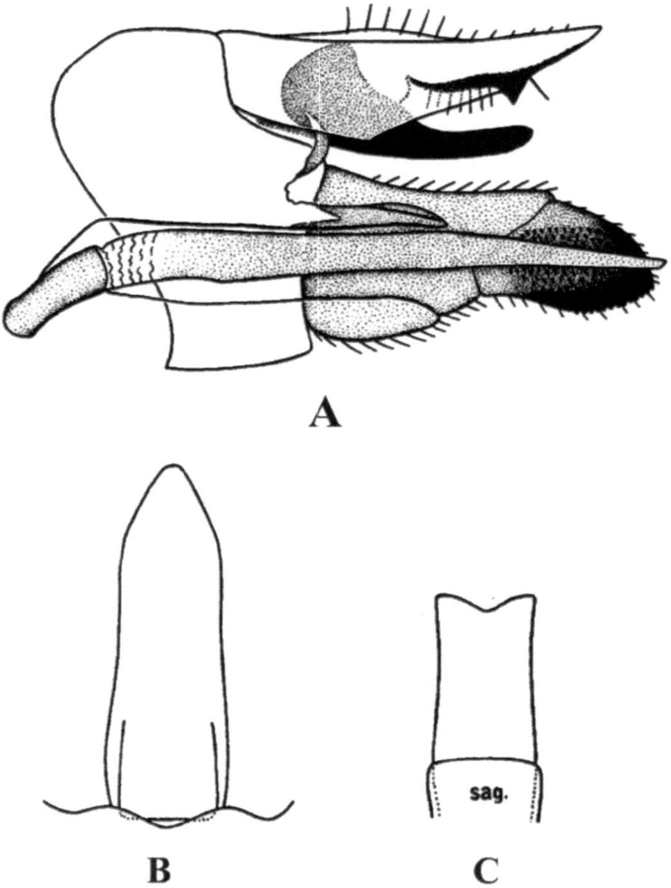

Figure 7.92 *Rhyacophila lhabu* (♂): (A) Lateral view of male genitalia; (B) Dorsal view; (C) Dorsal view of anal sclerite (from Schmid, 1970).

Holotype: INDIA: Manipur, Loni, 4300–5500 ft., 16-VI-1960.

Holotype depository: ROM, Canada.

Distribution: Vietnam, India (Manipur).

84. *Rhyacophila lhadzongpa* Schmid

Rhyacophila lhadzongpa, Schmid, 1970: 140

Adult: Head, maxillary palp, prothorax, abdomen and wings brown; scape of antenna black; pedicel brown and flagellum with alternating light and dark brown bands; mesothorax and metathorax dark brown; legs orangish brown, fore and mid legs with brown markings.

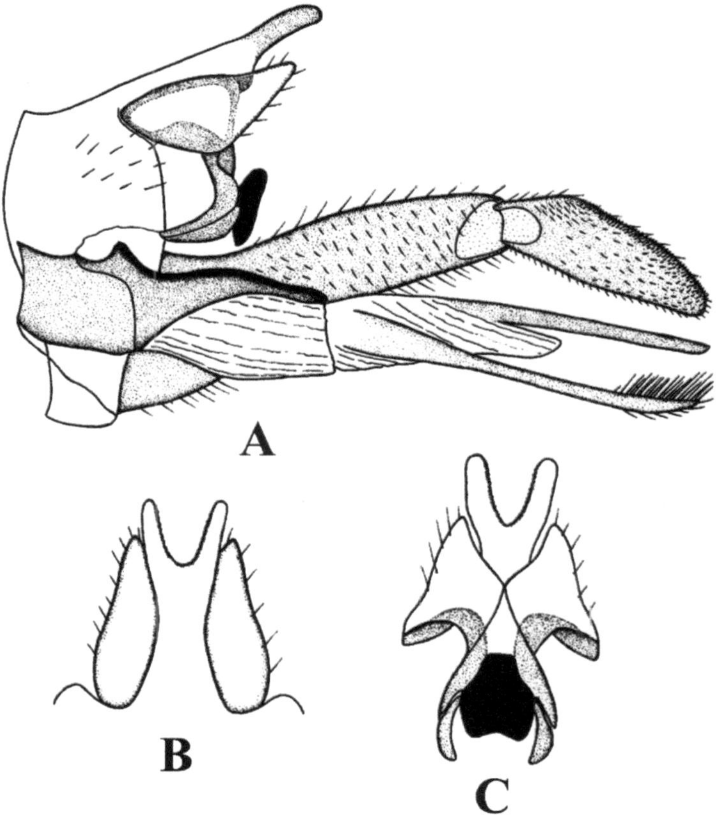

Figure 7.93 *Rhyacophila lhadzongpa* (♂): (A) Lateral view of male genitalia; (B) Dorsal view; (C) Front view of Xth segment (from Schmid, 1970).

Genitalia (♂): (Figs. 7.93A–C). Apical dorsal lobe of segment IX slender and deeply indented. Anal sclerites fused into an odd plate. Inferior appendages strongly elongated and markedly constricted: the second joint is long triangular and shows a slightly marked upper basal angle, although in general slightly deeper. Simple and Spiniferous Parameres.

Genitalia (♀): (Figs. 7.94A–E). Tergum and sternum of segment VII almost of the same length. Segment VIII smaller than VII segment. Dorsal apodeme extending anterad up to posterior margin of segment VII whereas basal apodeme near to anterior margin of segment VII and posteriorly with oval patch. Tergum of segment VIII truncate and sternum with V shape indentation and edges are rounded. Posterior process of vaginal apparatus sword shaped. Intermediate process curved and processus spermathecae small in size and roughly rectangular in shape.

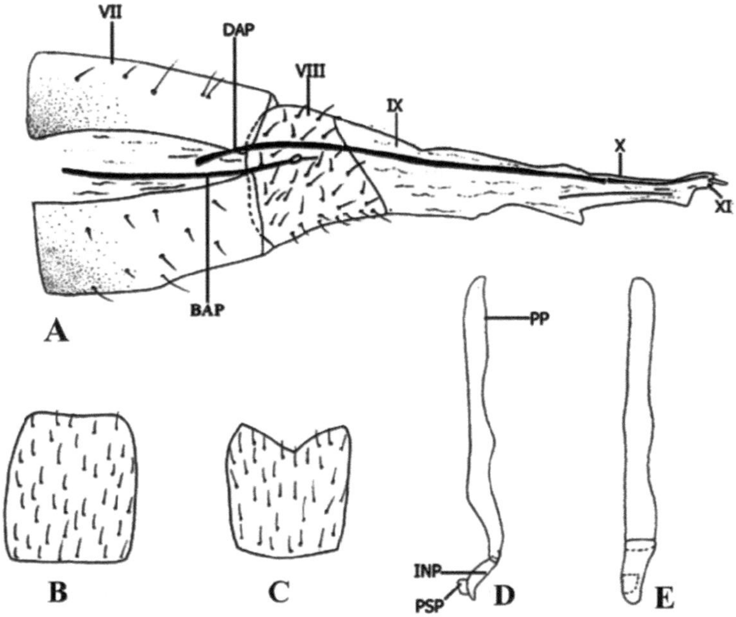

Figure 7.94 *Rhyacophila lhadzongpa* (♀): (A) Lateral view of female genitalia; (B) Dorsal view of segment VIII; (C) Ventral view of segment VIII; (D) Lateral view of vaginal apparatus; (E) Ventral view of vaginal apparatus.

Holotype: INDIA: Assam; NEFA, Kameng Frontier Division, Rupa, 5500–8000 ft., 1-IV-1961.

Holotype depository: ROM, Canada.

Additional material examined: INDIA: Jammu & Kashmir: Sonmarg, 3000 m, 26-VI-2011, 18 ♀♀; Himachal Pradesh: Dalhousie, 2000 m, 11-VIII-2011, 1♀. (Coll. Parey & Pandhar).

Distribution: Pakistan/India (Arunachal, Jammu and Kashmir, Himachal Pradesh).

85. *Rhyacophila lhakpa* Schmid

Rhyacophila lhakpa, Schmid, 1970: 174–175

Adult: Entire body dark brown except head, thorax and abdomen which are blackish in color. Length of antenna 5.33 mm and forewing 6.79 mm. In forewing fork 2nd arise earlier than 1st.

Genitalia (♂): (Figs. 7.95A–C). In dorsal view anal sclerite indented and apices of preanal appendage pointed. In lateral view tergum of segment IX longer than sternum. Preanal appendage smaller than anal sclerite. Dorsal appendage of phallic apparatus rounded, phallicata broader at base and ventral lobe of phallicata expanded. Paramere curved at middle. Basal segment of inferior appendage $1^{1/2}$ times broader than terminal segment of inferior appendage. Terminal segment of inferior appendage with thick sclerotization at apical margin. In ventral view terminal segment of inferior appendage rounded apically.

Genitalia (♀): Not available.

Holotype: INDIA: Uttarakhand; Pauri Garhwal, Hanuman Chatti, 7200–9000 ft., 17-VI-1958.

Holotype depository: ROM, Canada.

Additional material examined: INDIA: Himachal Pradesh: Kothi, 1900 m, 05-VII-2009; Barot, 2500 m, 06-VI-2010; Uttarakhand: Auli, 2900 m, 16-VI-2011. (Coll. Parey & Pandhar).

Distribution: India (Uttarakhand, Himachal Pradesh).

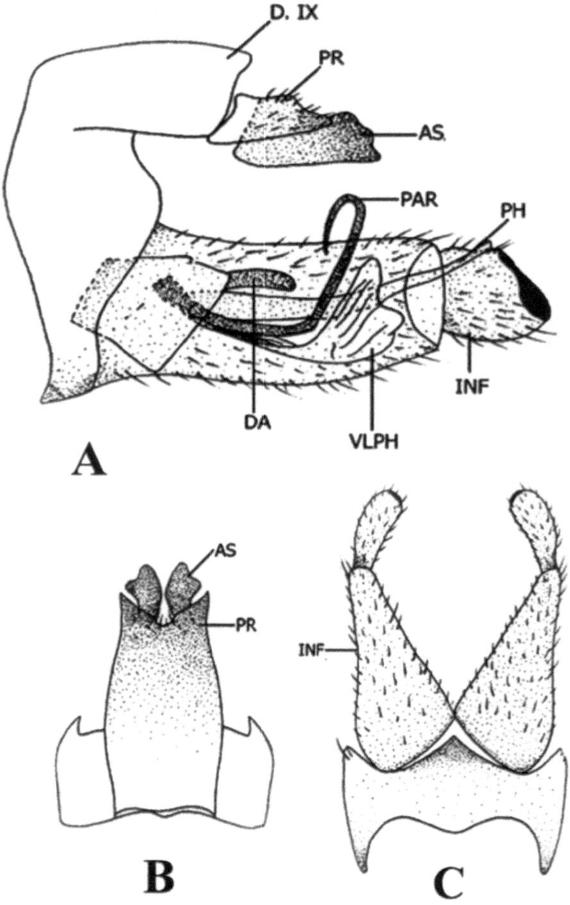

Figure 7.95 *Rhyacophila lhakpa* (♂): (A) Lateral view of male genitalia; (B) Dorsal view; (C) Ventral view.

86. *Rhyacophila lhopa* Schmid

Rhyacophila lhopa, Schmid, 1970: 197–198

Adult: Upper body dark brown, with concolorous hairs. Pleura and legs light reddish yellow. Narrow wings: the front ones are black-brown, with some very indistinct yellowish patterns. Wingspan: 16–22 mm.

Genitalia (♂): (Figs. 7.96A–C). Small preanal appendages. Internal face of the first segment of the inferior appendages with a strong bead extending the phallus. Second segment of inferior appendage with its outer face strongly hairy at its base and along its lower edge; the middle and upper portions are glabrous, strongly concave and drawn into a long, slender spine, arching outwards, overhanging the concavity and arising from the upper basal angle of the segment

Genitalia (♀): Not available.

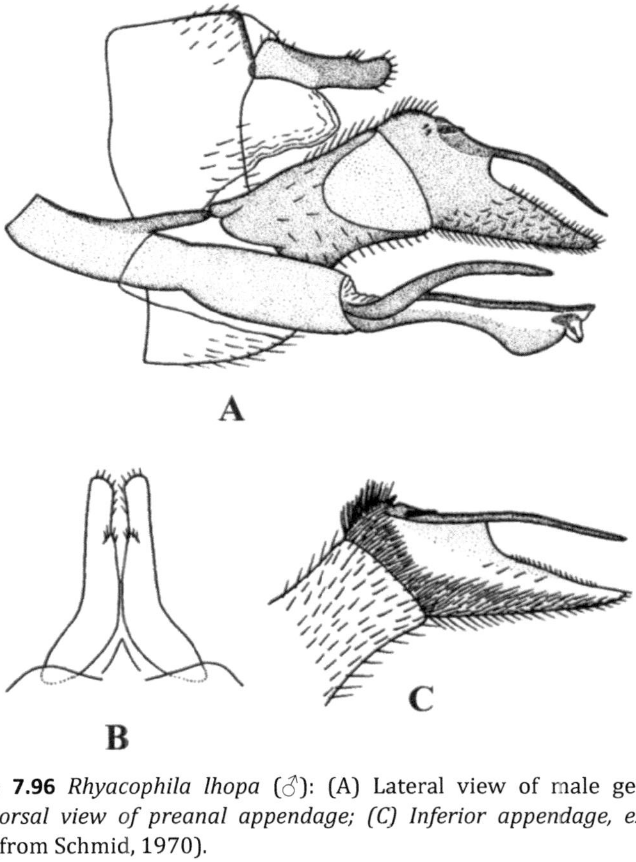

Figure 7.96 *Rhyacophila lhopa* (♂): (A) Lateral view of male genitalia; (B) *Dorsal view of preanal appendage; (C) Inferior appendage, external view* (from Schmid, 1970).

Holotype: INDIA: Madras, Kodaikanal, 3500–7300 ft., 5-12-XII-1958.

Holotype depository: ROM, Canada.

Distribution: India (Chennai).

87. *Rhyacophila lobsang* Schmid

Rhyacophila lobsang, Schmid, 1970: 146

Adult: Top of the head red, with dark golden hair. Pleuron and legs red. Forewings dark brown, with small inconspicuous silvery macules. Wingspan: 12.5–14 mm.

Genitalia (♂): (Figs. 7.97A–D). Xth segment with its rounded dorsal part; seen from the front, it shows two edges in arcs of a circle open laterally and forming an approximate X. Second segment of the inferior appendages very slightly indented. Long phallic apparatus; phallus thick, but with its lower terminal branch very long and slender; ventral plate with a weak triangular apical indentation.

Genitalia (♀): Not available.

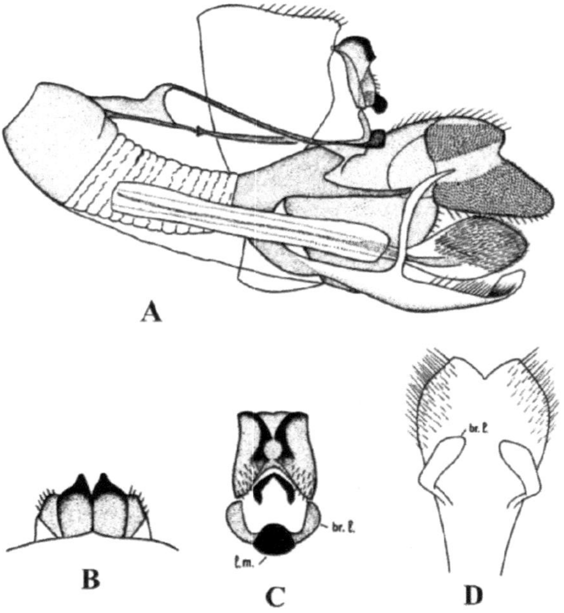

Figure 7.97 *Rhyacophila lobsang* (♂): (A) Lateral view of male genitalia; (B) Dorsal view of Xth segment; (C) Front view of Xth segment; (D) Ventral plate of phallus, dorsal view (from Schmid, 1970).

Holotype: INDIA: Madras, Meppadi, 1000–3000 ft., 9-I-1959.

Holotype depository: ROM, Canada.

Distribution: India (Chennai).

88. *Rhyacophila lonpo* Schmid

Rhyacophila lonpo, Schmid, 1970: 201

Adult: Dorsal side of the body reddish-brown, with golden hair. Pleura and legs reddish yellow. Moderately broad wings: the front ones are russet brown. Wingspan: 19–25 mm.

Genitalia (♂): (Figs. 7.98A–C). IXth segment strongly narrowed laterally by a wide indentation and present as a narrow band. Dorsal side of Xth segment without inferior apical angle; seen from above, it is regularly rectangular at its apical part and has a slightly marked median indentation. Dorsal appendage of the phallic apparatus very high, short, very convex dorsally at its base and depressed laterally before its extremity; apex of parameres somewhat membranous and armed with a helically twisted row of spines.

Genitalia (♀): Not available.

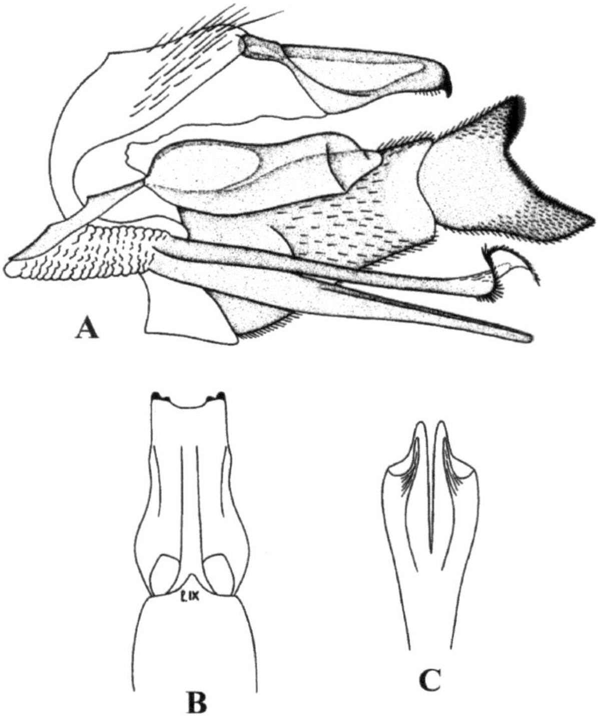

Figure 7.98 *Rhyacophila lonpo* (♂): (A) Lateral view of male genitalia; (B) Dorsal view; (C) Phallic apparatus, dorsal view (from Schmid, 1970).

Holotype: INDIA: Assam; NEFA, Kameng Frontier Division, Shergaon, 5300–7000 ft., 30-III-1961.

Holotype depository: ROM, Canada.

Distribution: India (Arunachal Pradesh).

89. *Rhyacophila maitripa* Schmid

Rhyacophila maitripa, Schmid, 1970: 211–212

Adult: Body dark brown, with dark but distinct mid-dorsal line. Genital parts dark in color. Very wide wings: the front ones do not have the strongly contrasting and mottled coloring of the other species of the group but are very dark and relatively uniform; they are dotted with thin golden macules not very clear, the light band of the apical area is often absent, but the post-costal area has a wide dark golden band, quite clearly visible. The dark spot of the subthyroid cell is inconspicuous. Wingspan 20–30 mm.

Genitalia (♂): (Figs. 7.99A–B). Apical dorsal lobe of segment IX very large, usually, but not always, exceeding dorsal apical angle of second segment of inferior appendages. The first segment

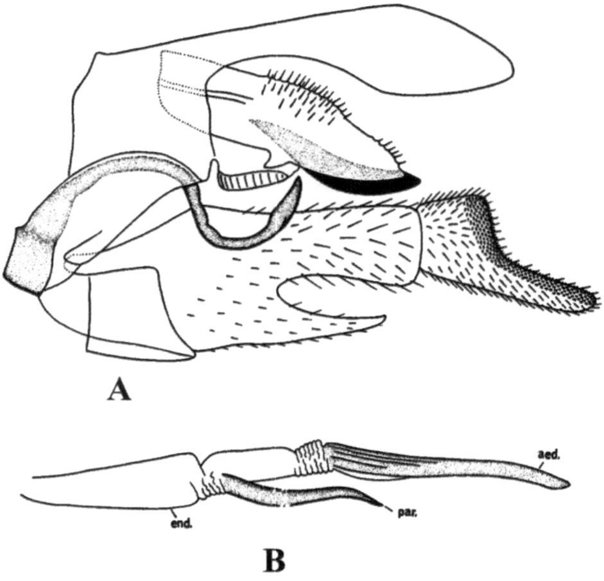

Figure 7.99 *Rhyacophila maitripa* (♂): (A) Lateral view of male genitalia; (B) *Lateral view of phallic apparatus* (from Schmid, 1970).

of the latter is longer and has a larger lower indentation, in a more regular oval, isolating a finer point. Phallus moderately long and fairly regular in thickness; parameres spiniform, thick, sinuous, distinctly shorter than the phallus and inserted on the endotheca long before the base of the phallus.

Genitalia (♀): (Figs. 7.100A–B). VIIIth segment a little more sclerotized than in the other species of the group and terminated in two large apical lobes, free but not very prominent.

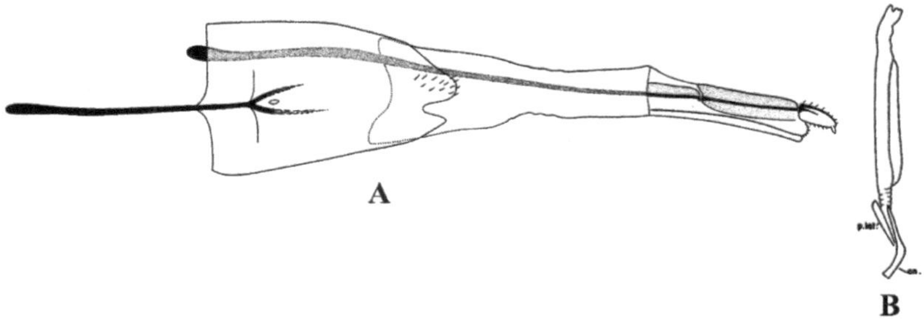

Figure 7.100 *Rhyacophila maitripa* (♀): (A) Lateral view of female genitalia; (B) *Lateral view of copulatory bursa* (from Schmid, 1970).

Holotype: INDIA: Assam; NEFA, Kameng Frontier Division, Jhum La, 6800–9000 ft., 15-22-IX-1961.

Holotype depository: ROM, Canada.

Distribution: India (Assam, Arunachal Pradesh).

90. *Rhyacophila manipuri* Schmid

Rhyacophila manipuri, Schmid, 1970: 140

Adult: Dorsal side of the body dark brown, with concolorous hairs. Red pleura; reddish-brown legs. Forewings blackish brown, with a few golden hairs. Wingspan: 15–20 mm.

Genitalia (♂): (Figs. 7.101A–D). Genitalia with all appendages remarkably short and thick. Apical dorsal lobe of the Xth segment not very indented at its extremity. Apex of the basal segment of the inferior appendages very thick and with a simple, but very pronounced internal relief. Parameres thick, strongly twisted and terminated in two unequal branches.

Genitalia (♀): Not available.

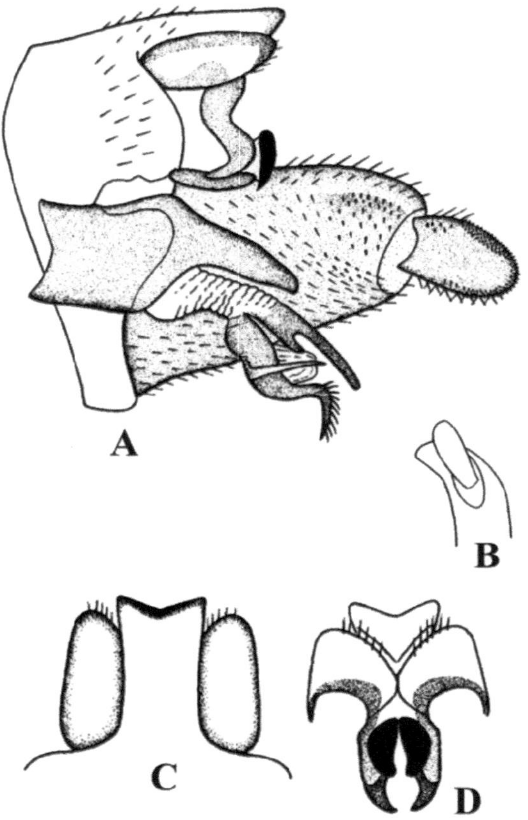

Figure 7.101 *Rhyacophila manipuri* (♂): (A) Lateral view of male genitalia; (B) Apex of inferior appendage, dorsal; (C) Dorsal view of IXth and Xth segment; (D) Front view of Xth segment (from Schmid, 1970).

Holotype: INDIA: Manipur; Marou, 3000–4500 ft., 14-VIII-1960.

Holotype depository: ROM, Canada.

Distribution: India (Manipur).

91. *Rhyacophila manlungpa* Schmid

Rhyacophila manlungpa, Schmid, 1970: 155

Adult: Top of the body dark brown, with concolorous hairs. Pleura reddish-brown, pale-yellow legs. Forewings blackish brown, with a few golden hairs. Wingspan: 16–22 mm.

Genitalia (♂): (Figs. 7.102A–B). Horizontal part of Xth segment long and strong; side faces of the segment very high; vertical part without upper point, but slightly concave. Second segment of the inferior appendages truncated, very obliquely and very slightly indented at its end. This species is easily recognized by its remarkably tall X^{th} segment and the vertical portion of the Xth segment armed with a single lower spike (Schmid, 1970).

Genitalia (♀): Not available.

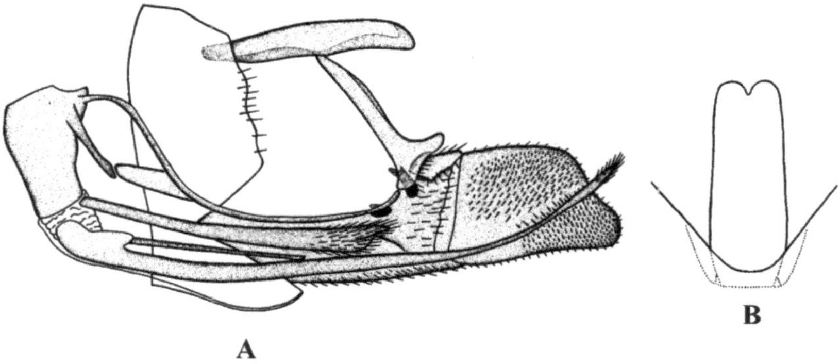

Figure 7.102 *Rhyacophila manlungpa* (♂): (A) Lateral view of male genitalia; (B) Dorsal view of Xth segment (from Schmid, 1970).

Holotype: INDIA: Assam: NEFA, Kameng Frontier Division; Arunachal Pradesh: Talung Dzong, 6000–8000 ft., 12-V-1961.

Holotype depository: ROM, Canada.

Distribution: India (Assam, Arunachal Pradesh).

92. *Rhyacophila marpa* Schmid

Rhyacophila marpa, Schmid, 1970: 177

Adult: Dorsal side of the body brown black, with black hairs. Dark reddish-brown pleura and legs. Bright orange abdomen. Uniformly black wings: the front ones carry at their extremity a small desclerotized zone, with not very clear outlines, covered with gray hairs. In case of females, there is a half-crescent white spot at the pterostigma. Wingspan: 14–15 mm.

Genitalia (♂): (Figs. 7.103A–D). Apical dorsal lobe of segment IX very high at its end and forming two strongly sclerotized lobes. Preanal appendages oval. Anal sclerites small, long ovate and half-hidden behind the preanal appendages. Phallic apparatus small and elongated; ventral plate of the aedeagus forming two small, parallel, strongly sclerotized lobes. Second segment of the inferior appendages widened at its extremity, and with its apical border slightly depressed.

Genitalia (♀): Not available.

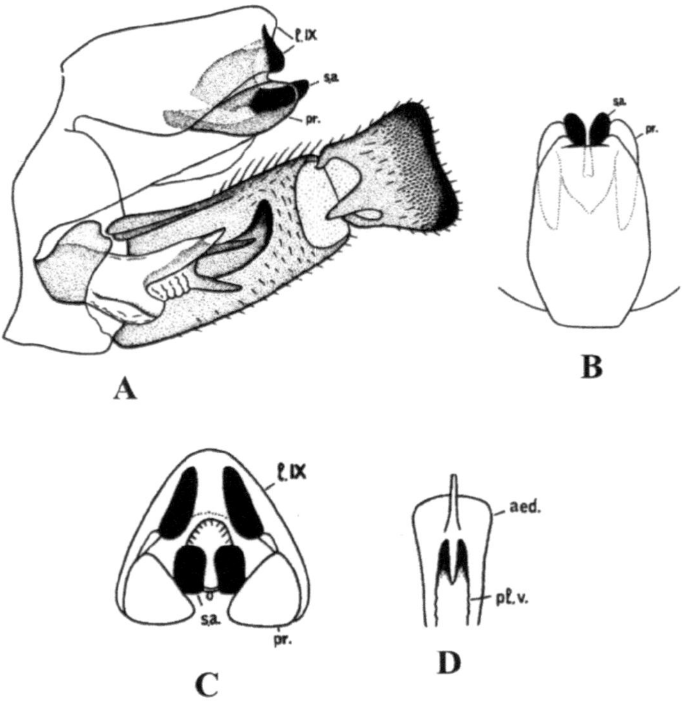

Figure 7.103 *Rhyacophila marpa* (♂): (A) Lateral view of male genitalia; (B) *Dorsal view; (C) Front view of* X^{th} *segment; (D) Phallic apparatus, dorsal view* (from Schmid, 1970).

Holotype: INDIA: Assam: NEFA, Kameng Frontier Division; Arunachal Pradesh: Chug, 7200 ft., 25-31-VII-1961.

Holotype depository: ROM, Canada.

Distribution: India (Assam, Arunachal Pradesh).

93. *Rhyacophila milarepa* Schmid

Rhyacophila milarepa, Schmid, 1970: 193

Adult: Entire body brown except labial palp which is pale brown. Length of antenna 8.24 mm and forewing 9.7 mm. In forewing fork 2nd arise slightly earlier than 1st.

Genitalia (♂): (Figs. 7.104A–C). In dorsal view, anal sclerite apically rounded and anal sclerite and preanal appendage fused. In lateral view tergum of segment IX same in length as of sternum. Dorsal apical lobe of segment IX membranous. Preanal appendages convexly raised near apex and anal sclerite rounded apically. Phallicata broader at base and truncate apically. Paramere also broader at base and gradually go on decreasing up to apex. First segment of the inferior appendage slightly indented at middle. From the first segment, arise long tuft of setae. Inferior lobe of terminal segment of inferior appendage with short spinules. In ventral view basal segment of inferior appendage is with concavity at the middle.

Genitalia (♀): Not available.

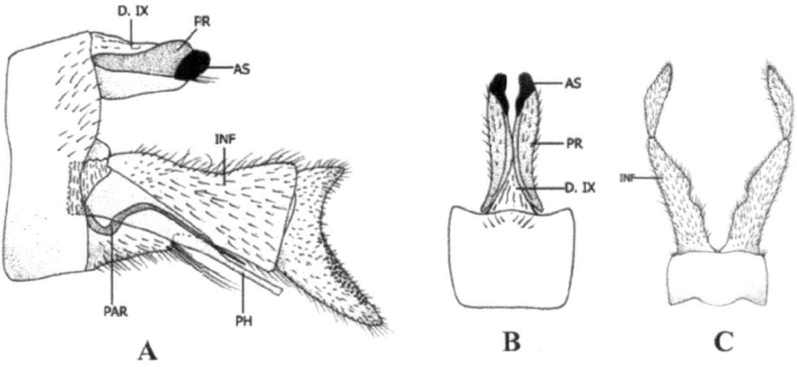

Figure 7.104 *Rhyacophila milarepa* (♂): (A) Male genitalia lateral view; (B) Dorsal view; (C) Ventral view.

Holotype: INDIA: Assam: NEFA, Kameng Frontier Division; Arunachal Pradesh: Bomdi La, 13-21-VI-1961.

Holotype depository: ROM, Canada.

Additional material examined: INDIA: Arunachal Pradesh: Bomdila, 2400 m, 03-X-2010, 9 ♂♂. (Coll. Parey & Pandhar).

Distribution: India (Assam, Arunachal Pradesh).

94. *Rhyacophila mishmica* Kimmins

Rhyacophila milarepa, Kimmins, 1953: 517–519

Adult: Ground color of body fuscous, with fuscous and golden hairs. Antennae fulvous, with fuscous annulation on each segment. Wings elongate, pubescence brown, with numerous golden irrorations. Membrane smoky brown, with pale irrorations.

Genitalia (♂): (Figs. 7.105A–C). IXth segment with lateral margins deeply excised and sinuous, ventral margin rounded. Dorsal process large, slightly tapering, fused at its apex to the Xth tergite. The latter consists of a pair of elongate plates, fused along their upper margins almost to the apex, forming a long, narrow hood with a blackened, excised apex. From the side the tergite ends in a small claw. Xth sternite membranous. Phallus long, slender, straight, in side view narrowed apically. Sheaths almost as long as phallus, straight, slender, apices membranous, dilated and then tapering and armed with a tuft of setae. Inferior appendage long, moderately broad at base, very slightly constricted beyond middle in lateral view. Terminal segment short, dilating to a widely excised apex, lower branch larger and longer than upper, and more acute.

Genitalia (♀): Not available.

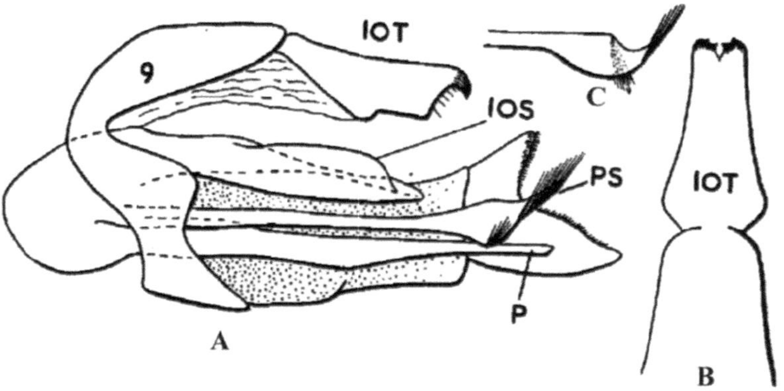

Figure 7.105 *Rhyacophila mishmica* (♂): (A) Lateral view of male genitalia; (B) Dorsal view of Xth tergite; (C) Apex of phallus sheath, dorsal (from Kimmins, 1953).

Holotype: INDIA: Assam: Mishmi Hills; Arunachal Pradesh: Delai Valley, Che Che, 22-X1-1936.

Holotype depository: BMNH, London.

Distribution: India (Assam, Arunachal Pradesh).

95. *Rhyacophila monyulpa* Schmid

Rhyacophila monyulpa, Schmid, 1970: 165

Adult: Top of the body black, with concolorous hairiness. Fore and median pleura, corresponding legs and hind tarsi dark rufous brown. Abdomen, posterior pleura, femur and base of the corresponding tibia pinkish orange. Gray black forewings, with clearly visible veins and a translucent zone, with indistinct outlines, at the apex of the apical cells. Wingspan: 19–24 mm.

Genitalia (♂): (Figs. 7.106A–B). Dorsal lobe very long, narrow gradually tapering to its end, which is bilobed. Very long, thin and also bilobed anal scleritis. First segment of the inferior appendages with strong internal relief; regularly subcircular second segment.

Genitalia (♀): Not available.

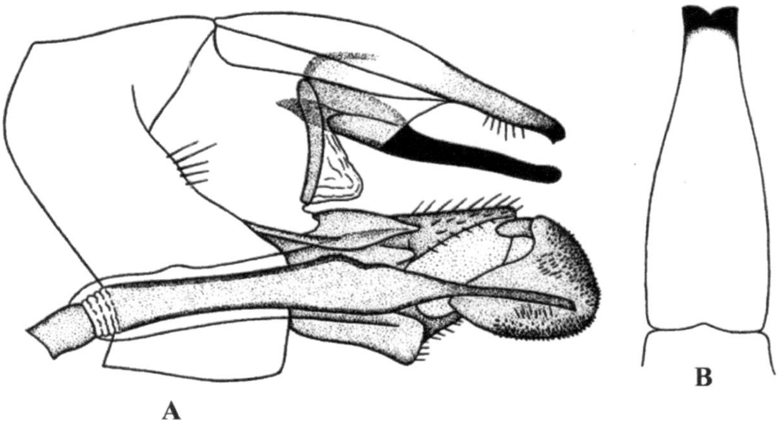

Figure 7.106 *Rhyacophila monyulpa* (♂): (A) Lateral view of male genitalia; (B) Dorsal view (from Schmid, 1970).

Holotype: INDIA: Assam: NEFA, Kameng Frontier Division; Arunachal Pradesh: Chug, 5200–6500 ft., 17-IV-1961.

Holotype depository: ROM, Canada.

Distribution: India (Assam, Arunachal Pradesh).

96. *Rhyacophila muktepa* Schmid

Rhyacophila muktepa, Schmid, 1970: 188–189

Adult: Dorsal side of the body reddish-brown, with concolorous hairiness. Rufous yellow legs. Uniformly brown wings. Wingspan: 11.5–20 mm.

Genitalia (♂): (Figs. 7.107A–B). Sub-quadrangular preanal appendages, with their angles rounded. Anal sclerites appearing triangular, seen laterally, with their upper edge oblique downwards. Phallus composed of two sub-equal branches. Second segment of the inferior appendages triangular, with its upper edge domed.

Genitalia (♀): Not available.

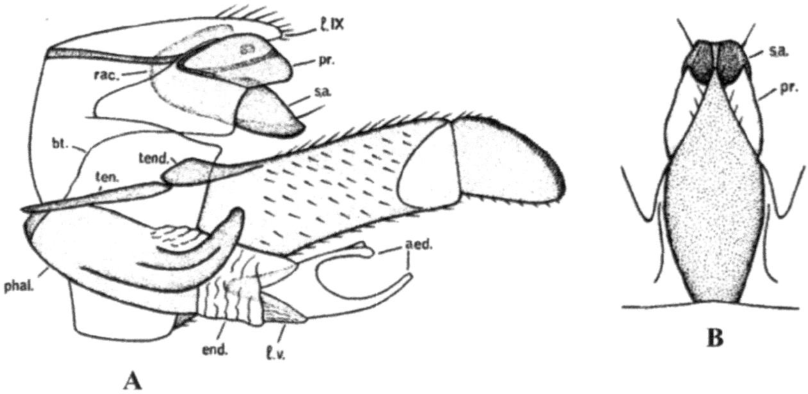

Figure 7.107 *Rhyacophila muktepa* (♂): (A) Lateral view of male genitalia; (B) Dorsal view (from Schmid, 1970).

Holotype: INDIA: Assam: United Jaintia and Khasia Hills; Rumkheng, 1800–5500 ft., 20-23-III-1960.

Holotype depository: ROM, Canada.

Distribution: Thailand, India (Meghalaya).

97. *Rhyacophila nabochepa* Schmid

Rhyacophila nabochepa, Schmid, 1970: 178–179

Adult: This species shows quite variable coloration. Most specimens have all-black bodies and wings, with orange abdomens. But some specimens have discolored wings. Hairs of the top of the body black. Wingspan: 13–19 mm (Schmid, 1970).

Genitalia (♂): (Figs. 7.108A–B). Apical dorsal lobe of segment IX not very flared at its extremity, forming two lateral lobes of quite variable shape. Preanal appendages in parallelograms. Anal sclerites in hooks curved upwards. Dorsal appendages of the phallic apparatus strong and angular; the two branches of the phallus are slender cylinders widely separated from each other. Second segment of the inferior appendages regularly triangular.

Genitalia (♀): Not available.

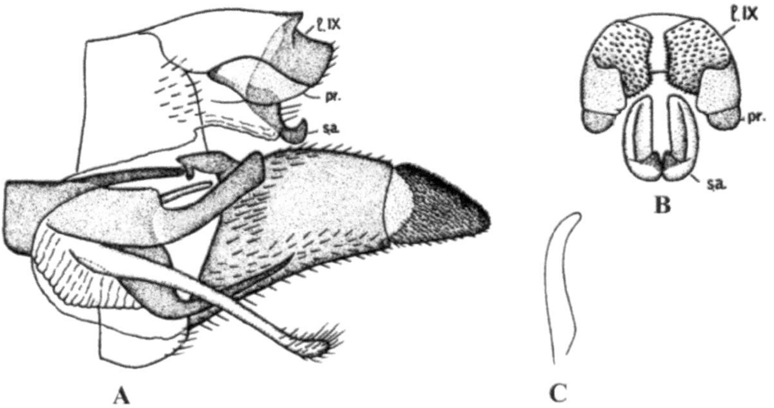

Figure 7.108 *Rhyacophila nabochepa* (♂): (A) Lateral view of male genitalia; (B) Front view of Xth segment; (C) Phallic apparatus, dorsal (from Schmid, 1970).

Holotype: INDIA: Western Bengal, Dilpa, 5500–10000 ft., 21-III-1959.

Holotype depository: ROM, Canada.

Distribution: Thailand, India (West Bengal).

98. *Rhyacophila naga* Schmid

Rhyacophila naga, Schmid, 1970: 155

Adult: Upper body dark brown, with colored hairs. Legs and pleura yellow-brown. Wings uniformly blackish. Wingspan 18–22 mm.

Genitalia (♂): (Figs. 7.109A–D). Vertical part of the Xth segment with two apical points; upper point a little longer than the horizontal part of the segment and shortly bifid; lower tip half smaller and entire. The two segments of the inferior appendages have their edges aligned; second segment higher than long, bearing an apical indentation isolating two unequal and regularly rounded lobes.

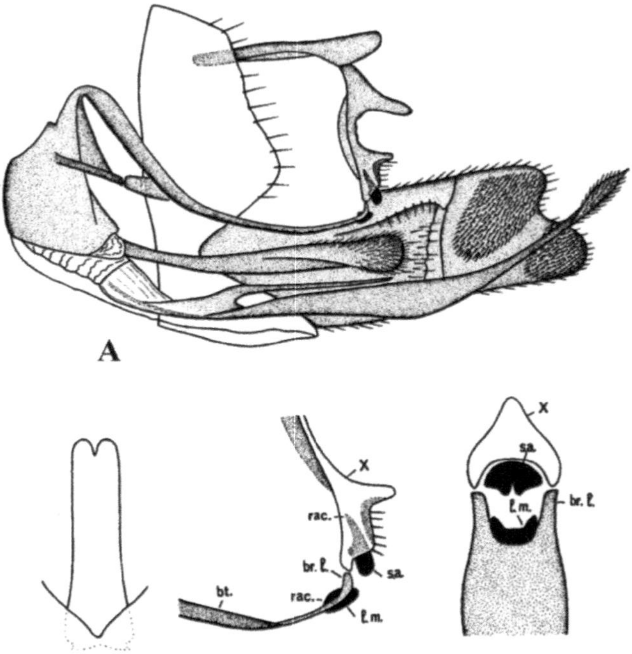

Figure 7.109 *Rhyacophila naga* (♂): (A) Lateral view of male genitalia; (B) Dorsal view of Xth segment; (C) Lower part of Xth segment, lateral; (D) Front view (from Schmid, 1970).

Genitalia (♀): (Fig. 7.110). Terminal part of the chitinous copulatory bursa on its apical half regularly ogival.

Holotype: INDIA: Manipur, Hkayam Boum, 3800–8500 ft., 20-23-VI-1960.

Holotype depository: ROM, Canada.

Distribution: India (Manipur).

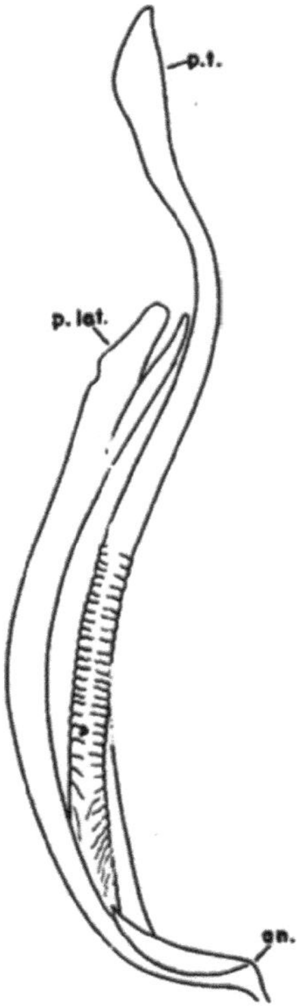

Figure 7.110 *Rhyacophila naga* (♀): Copulatory bursa (from Schmid, 1970).

99. *Rhyacophila nagongpa* Schmid

Rhyacophila nagongpa, Schmid, 1970: 140–141

Adult: Dorsal side of the body dark brown, with concolorous hairiness. Red pleura; brown legs. Forewings dark brown, dotted with golden macules, fine, few and inconspicuous. Wingspan: 15–20 mm.

Genitalia (♂): (Figs. 7.111A–D). Apical dorsal lobe of segment IX almost not indented at its tip. Laterally, preanal appendages appearing oval. Inferior appendages massive, with the apex of the first segment thick and the second segment oval and relatively large. Dorsal appendage of phallic apparatus terminated in two slender apical points; simple parameres.

Genitalia (♀): Not available.

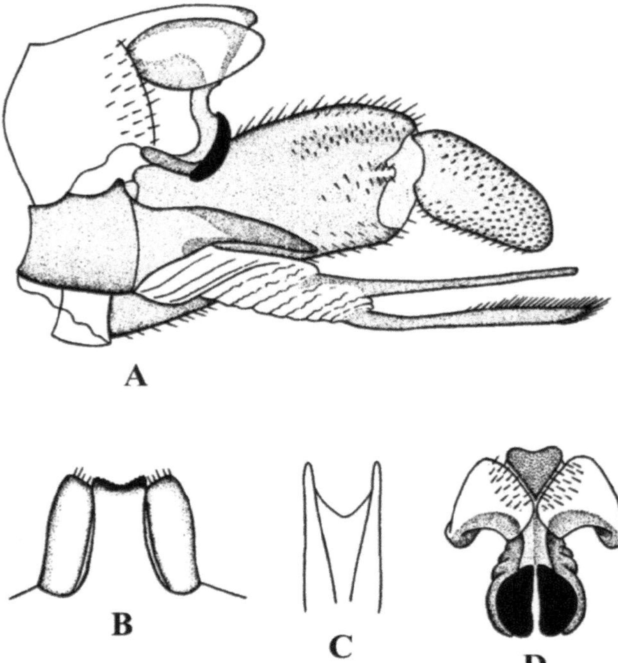

Figure 7.111 *Rhyacophila nagongpa* (♂): (A) Lateral view of male genitalia; (B) Dorsal view; (C) Apex of phallic apparatus, dorsal; (D) Front view of Xth segment (from Schmid, 1970).

Holotype: INDIA: Assam, United Jaintia and Khasi Hills, Muktapur, 11-12-XII-1959.

Holotype depository: ROM, Canada.

Distribution: India (Meghalaya).

100. *Rhyacophila nakpo* Schmid

Rhyacophila nakpo, Schmid, 1970: 165–166

Adult: Top of the body brown black, with concolorous hairiness. Reddish-brown pleura and legs. Wings uniformly black-brown. Wingspan: 21–22 mm.

Genitalia (♂): (Figs. 7.112A–C). Apical dorsal lobe of segment IX terminated in two acute points; well-distinct and complex-shaped preanal appendages at their extremity. Anal scleritis not very large, slightly bilobed and curved upwards. Sagittal appendage truncated oval. Inferior appendages in high relief, with the ventral edge of the first segment deeply incised and the second segment facing downwards and barely longer than high.

Genitalia (♀): Not available.

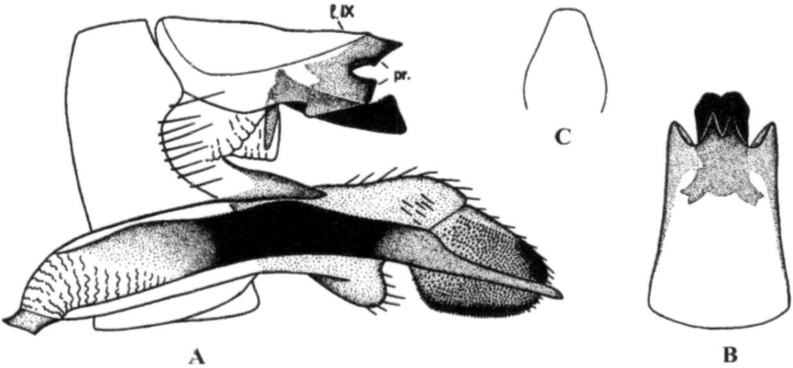

Figure 7.112 *Rhyacophila nakpo* (♂): (A) Lateral view of male genitalia; (B) Dorsal view; (C) sagittal appendage, dorsal (from Schmid, 1970).

Holotype: INDIA: Manipur; Sirohi Kashong, 6800–8500 ft., 8-VI-1960.

Holotype depository: ROM, Canada.

Distribution: India (Manipur).

101. *Rhyacophila namgyal* Schmid

Rhyacophila namgyal, Schmid, 1970: 146

Adult: Dorsal side of the body reddish-brown, with concolorous hairiness. Pleura and paws yellow. Forewings tinder golden, very dark, with golden and silver macules visible. Wingspan: 13–17.5 mm.

Genitalia (♂): (Figs. 7.113A–D). Xth segment slightly curved backwards at its apical dorsal part and armed with two caudal spines, directed downwards, and forming an arc of a circle. Second segment of the inferior appendages terminated in two lobes, the lower in a very elongated and horizontal oval. Phallus slender, with its well-marked dorsal branch.

Genitalia (♀): Not available.

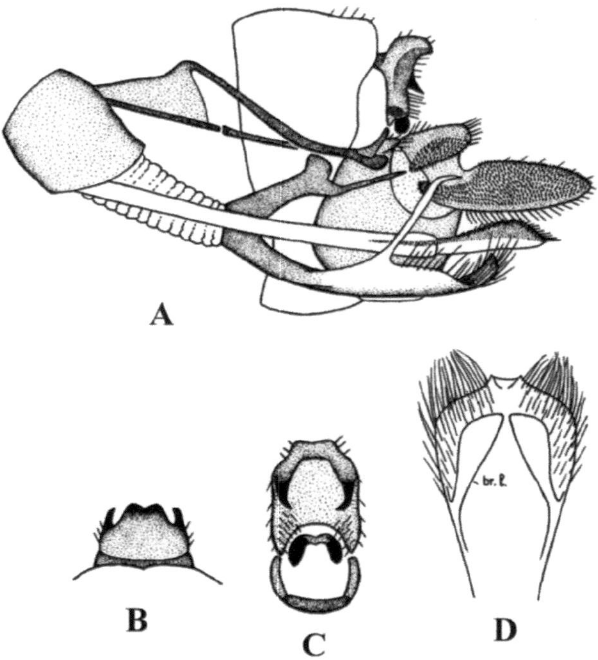

Figure 7.113 *Rhyacophila namgyal* (♂): (A) Lateral view of male genitalia; (B) Dorsal view of Xth segment; (C) Front view of Xth segment; (D) Ventral plate of phallus, dorsal view (from Schmid, 1970).

Holotype: INDIA: Assam; United Jaintia and Khasi Hills, Pynter, 250–4100 ft., 23-I-1960.

Holotype depository: ROM, Canada.

Distribution: India (Meghalaya).

102. *Rhyacophila narayani* Schmid

Rhyacophila narayani, Schmid, 1970: 198

Adult: Dorsal side of the body reddish yellow, with golden hairs. Pleura and legs light reddish yellow. Forewings brown, intensely and evenly riddled with small yellow macules. Abdominal tergites covered with very long fine hairs. Wingspan: 19–20 mm.

Genitalia (♂): (Figs. 7.114A–B). Vth sternite with a finger-like lobe along its upper edge. Preanal appendages entirely united to each other by a membrane, concave inwards and oblique downwards. Phallus short, high, laterally flattened except at its apical dorsal part which is spread out horizontally and ends in two very unequal branches. Second segment of the inferior appendages very small and oblique.

Genitalia (♀): Not available.

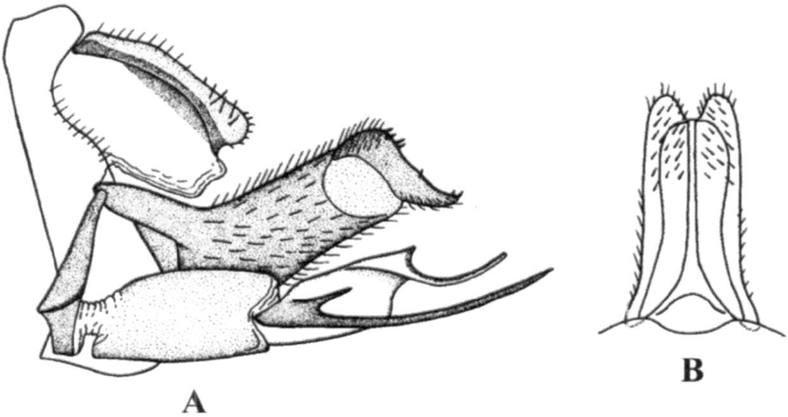

Figure 7.114 *Rhyacophila narayani* (♂): (A) Lateral view of male genitalia; (B) Dorsal view preanal appendage (from Schmid, 1970).

Holotype: INDIA: Kerala; Kandalur, 5500–6500 ft., 12-XII-1961.

Holotype depository: ROM, Canada.

Distribution: India (Kerala).

103. *Rhyacophila naviculata* Morton

Rhyacophila naviculata, Morton, 1900: 6–7

Adult: Anterior wings rather broad, apex parabolic; apical forks 1 and 2 sub-equal in length; radius not distinctly furcate. Pale yellowish with greyish reticulations. (Morton, 1900).

Genitalia (♂): (Figs. 7.115A–B). IXth segment produced in a deep, hooded dorsal process. The apex has a V-shaped excision from above, and shortly before the apex the hood is constricted in a narrow neck. Xth tergite small, consisting of a pair of plates hanging down from the margins of the dorsal process, and to their lower margin are attached a pair of rounded lobes linked basally by a Y-shaped structure. Xth sternite short, semi-annular. Basal sleeve of phallus with a strong process on its upper surface, stout basally, sinuous from the side and tapering to a slender spine-like apex. Inferior appendage long and slender, basal segment about three times as long as terminal, which is laterally compressed, elongate-oval, slightly upcurved and fringed with stout setae.

Genitalia (♀): Not available.

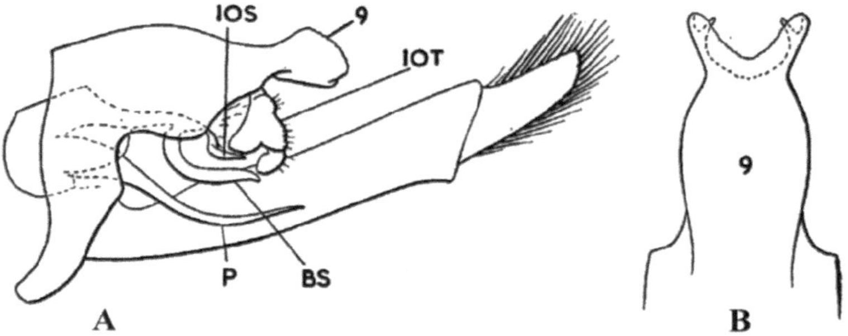

Figure 7.115 *Rhyacophila naviculata* (♂): (A) Lateral view of male genitalia; (B) Dorsal view IXth segment (from Kimmins, 1953).

Holotype: INDIA: Madras; Trichinopoly, 6-XII-1899.

Holotype depository: BMNH, London.

Holotype depository: BMNH, London.

Distribution: India (Tamil Nadu, Meghalaya).

104. ***Rhyacophila netongpa*** **Schmid**

Rhyacophila netongpa, Schmid, 1970: 194

Adult: Top of the body brown black, with concolorous hairiness. Reddish-brown pleura and legs. Wings uniformly black-brown. Wingspan: 23 mm.

Genitalia (♂): (Figs. 7.116A–B). Preanal appendages two and a half times longer than high, with their extremity appearing truncated very obliquely downwards, in lateral view; in dorsal view, they are fused to each other on their apical half. Phallus gradually tapering, slender and with a faint superior subapical angle; spiniform parameres, strongly contorted and armed with a series of thick spines along their lower edge. Second segment of the inferior appendages terminated in two rather broadly rounded lobes.

Genitalia (♀): Not available.

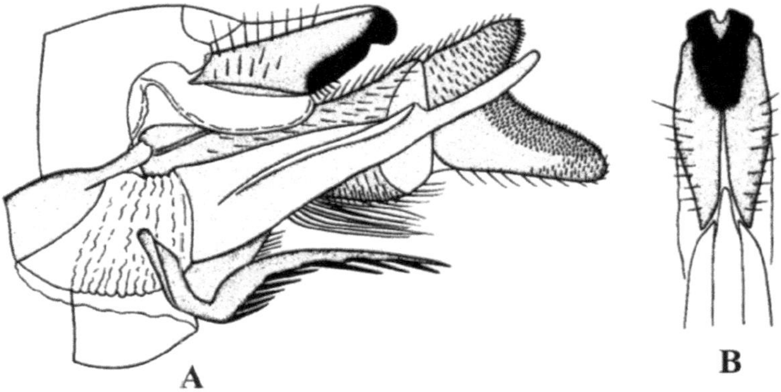

Figure 7.116 *Rhyacophila netongpa* (♂): (A) Male genitalia, lateral view; (B) Dorsal view of Xth segment (from Schmid, 1970).

Holotype: INDIA: Assam; NEFA, Kameng Frontier Division, Arunachal Pradesh; Jhum La, 31-V-1961.

Holotype depository: ROM, Canada.

Distribution: India (Assam, Arunachal Pradesh).

105. *Rhyacophila ngawang* Schmid

Rhyacophila ngawang, Schmid, 1970: 147

Adult: Dorsal side of the body dark reddish-brown, with concolorous hairiness. Pleura and paws red. Tinder golden forewings, very dark, with inconspicuous silver blotches. Wingspan: 13–14 mm.

Genitalia (♂): (Figs. 7.117A–D). Xth segment rather large and massive, appearing strongly and widely indented in the middle when viewed from above. Second segment of the inferior appendages terminated in two lobes of unequal length and furnished with two zones of elongated and parallel internal spinules. Phallus very thick and without dorsal branch.

Genitalia (♀): Not available.

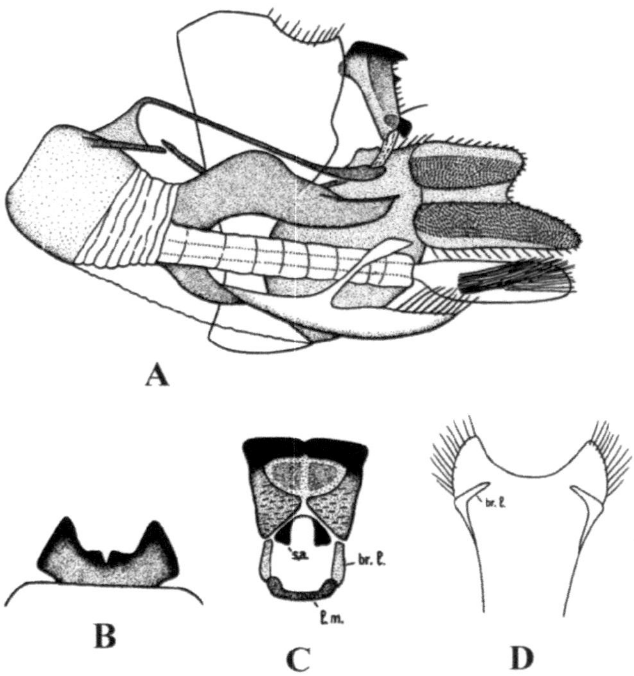

Figure 7.117 *Rhyacophila ngawang* (♂): (A) Lateral view of male genitalia; (B) Dorsal view of Xth segment; (C) Front view of Xth segment; (D) Ventral plate of phallus, dorsal view (from Schmid, 1970).

Holotype: INDIA: Assam, United Jaintia and Khasi Hills, Mawlang, 12-IV-1960.

Holotype depository: ROM, Canada.

Distribution: India (Meghalaya).

106. *Rhyacophila ngorpa* Schmid

Rhyacophila ngorpa, Schmid, 1970: 215

Adult: Entire body brown except maxillary and labial palpi which are pale brown. Length of antenna 8.73 mm and forewing 13.0 mm. In forewing forks 1st and 2nd arise at the same level.

Genitalia (♂): (Figs. 7.118A–C). In dorsal view apex of dorsal apical lobe of segment IX trifid, median lobe longer than lateral lobes. In lateral view segment IX much produced in long dorsal apical lobe of segment IX. Sternum of segment IX is much smaller than its tergum. Inferior appendages elongate and breadthness of basal segment almost same throughout. Terminal segment of inferior appendage short and deeply bifid, superior lobe shorter than inferior lobe. Segment X short, situated beneath the dorsal apical lobe of segment IX forming oblique plate. Anal sclerite bifid and slightly under the segment IX. Phallicata long, slender and rounded at apex. Paramere long and spiniform. In ventral view inferior lobes of terminal segment of inferior appendage crossing other.

Genitalia (♀): Not available.

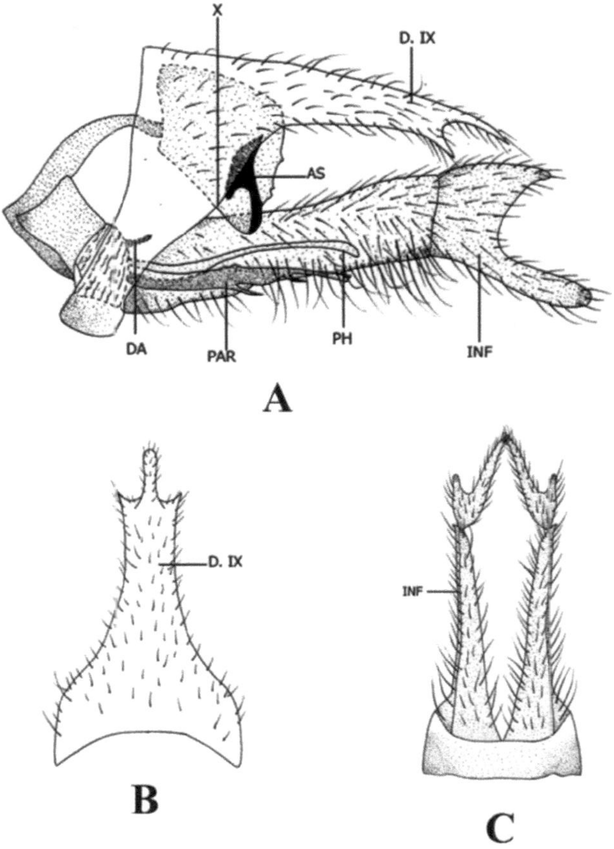

Figure 7.118 *Rhyacophila ngorpa* (♂): (A) Lateral view of male genitalia; (B) Dorsal view; (C) Ventral view.

Holotype: INDIA: West Bengal; Shepi, 23-III-1959.

Holotype depository: ROM, Canada.

Additional Material examined: INDIA: Sikkim: Golitar, 2200 m, 25-IV-2009. (Coll. Parey & Pandhar).

Distribution: Bhutan, Nepal, Tibet, India (West Bengal, Sikkim).

107. *Rhyacophila ngulpa* Schmid

Rhyacophila ngulpa, Schmid, 1970: 171

Adult: Head, mesothorax, metathorax and legs brown; antenna, maxillary palp, labial palp, prothorax and abdomen pale brown. Length of antenna 12.6 mm and forewing 15.2 mm. In forewing forks 1st and 2nd arise at the same level whereas in females, entire body brown; antenna with alternating light and dark brown bands; forewing with dark brown irrotations.

Genitalia (♂): (Figs. 7.119A–C). In dorsal view, dorsal apical lobe of segment IX trilobed, outer lobes long and round whereas median lobe is very smaller and triangular shaped. Anal sclerite hidden beneath dorsal apical lobe of segment IX. In lateral view, dorsal apical lobe of segment IX very large and apically widely excised. Inferior appendage long, basal segment narrows towards apex; terminal segment indented forming two unequal lobes and thickly spinose at apical margin. Preanal appendage long and narrow towards apex. Anal sclerite short and bilobed, external lobe less sclerotized. Phallicata with median superior spine, ventral lobe of phallicata robust, stick like and apically broad and curved downwards; paramere straight and apically with long spine. In ventral view basal segment with concavity at apex.

Genitalia (♀): (Figs. 7.120A–E). Laterally, VIIIth segment $1/3^{rd}$ in length as compared to VIIth segment and its dorsal margin almost half of the ventral margin. Posterolaterally, VIIIth segment with L shape indentation. Length of basal apodeme almost same as that of segment VII and it bifurcate posteriorly. Tergum of VIIIth segment with deep indentation and a deep parallel groove in its center. In ventral view segment VIII trilobed with median lobe further slightly notched. In lateral view, posterior process of vaginal apparatus rounded and intermediate process long.

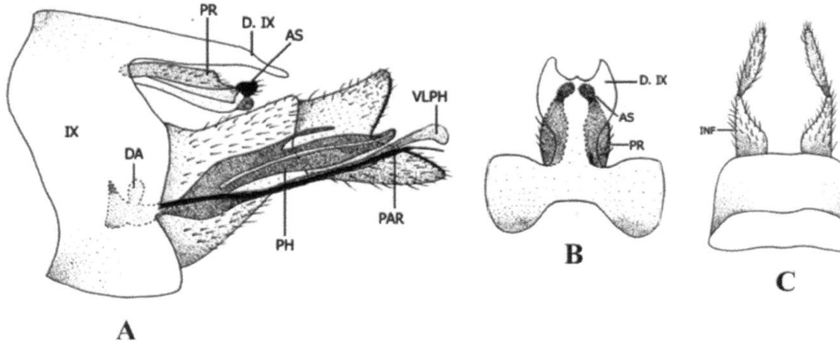

Figure 7.119 *Rhyacophila ngulpa* (♂): (A) Lateral view of male genitalia; (B) Dorsal view; (C) Ventral view.

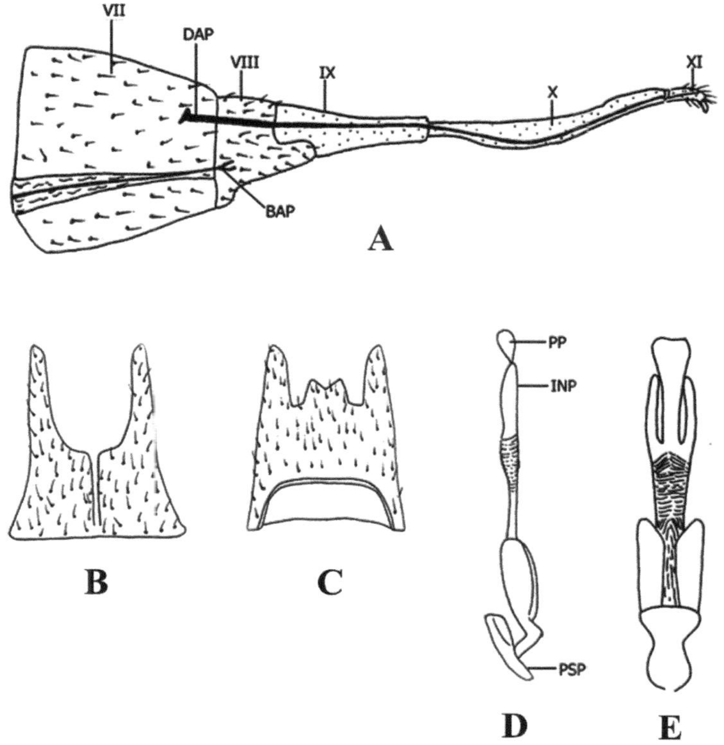

Figure 7.120 *Rhyacophila ngulpa* (♀): (A) Lateral view of female genitalia; (B) Dorsal view of segment VIII; (C) Ventral view of segment VIII; (D) Lateral view of vaginal apparatus; (E) Ventral view of vaginal apparatus.

Holotype: INDIA: Pauri Garhwal, Gangrea, 5800–11200 ft., 12-15-VI-1958.

Holotype depository: ROM, Canada.

Additional material examined: INDIA: Himachal Pradesh: Keylong, 3100 m, 10-VIII-2008; Dhorila, 3200 m, 16-VIII-2008; Chhatradu, 3600 m, 21-VIII-2008; Chandertal, 4300 m, 07-VII-2009; Uttarakhand: Gobinddham, 3100 m, 22-VI-2008; Badrinath, 3400 m, 24-VI-2008; Malarae, 3500 m, 19-VI-2011; Mana, 3200 m, 23-VI-2011; Mana, 3200 m, 21-VI-2012. (Coll. Parey & Pandhar).

Distribution: Nepal, India (Uttarakhand, Himachal Pradesh).

108. *Rhyacophila nigrorosea* Schmid

Rhyacophila nigrorosea, Schmid, 1959: 236

Adult: Head and thorax entirely black, as well as their appendages. Abdomen beautiful light orange-pink which colors even the genitalia; only the apex of the lower appendages is black. Wings entirely black; the hindwings are little lighter than the anterior ones and some females show a circular white patch at the pterostigma of the forewings.

Genitalia (♂): (Figs. 7.121A–C). Segment IX quite regularly narrow laterally and ventrally; dorsally, it extends into a large elongated lobe, domed and notched at its extremity. The Xth segment is reduced to its dorsal part which forms a set of small size, of complex structure, concave towards the top and prolonging the IXth segment. Short, massive lower appendages; very high basal segment; apical segment triangular and profusely furnished with internal apical spines. The phallic apparatus is very complex. Phallus in long thin tube and not very chitinous. Titillators in long cylindrical spines, curved upwards at right angles and ending in a bouquet of spines.

Genitalia (♀): (Figs. 7.122A–E). VIth segment with small spine ventrally and tergum of VII segment larger than its sternum. In lateral view segment VIII with deep indentation posteriorly. Basal apodeme extending anteriorly up to VI segment and with oval structure posteriorly but dorsal apodeme arise from segment VIII. Tergum of segment VIII with V shape indentation and sternum with deep notch. Posterior process of vaginal apparatus bilobed in ventral view.

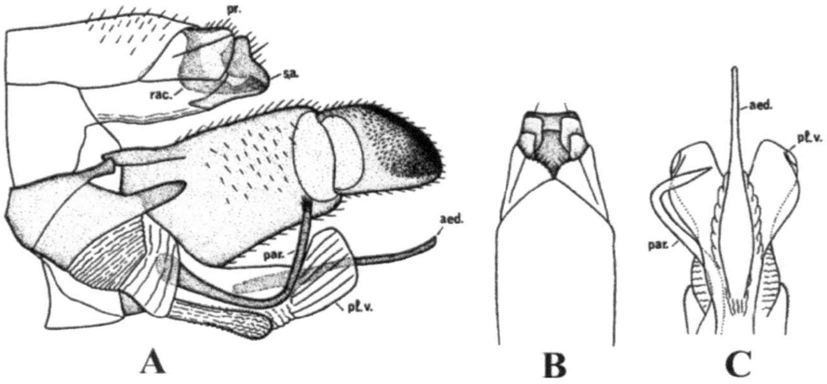

Figure 7.121 *Rhyacophila nigrorosea* (♂): (A) Lateral view of male genitalia; (B) Dorsal view; (C) Phallic apparatus, dorsal view (from Schmid, 1970).

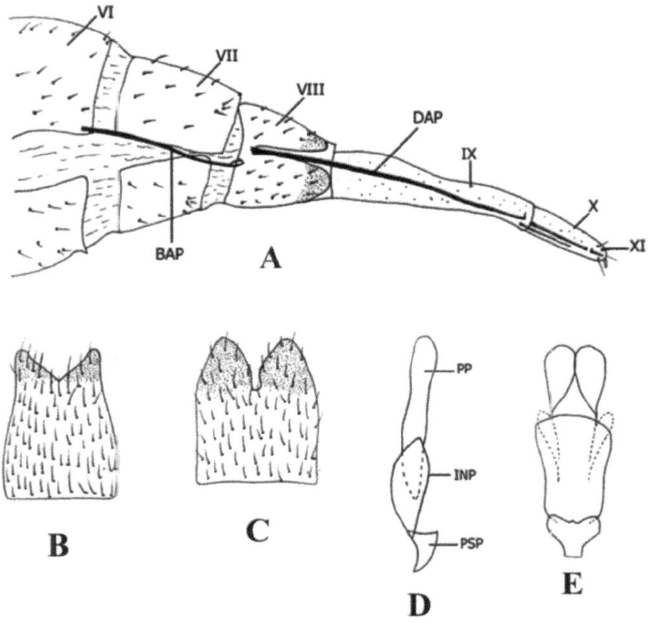

Figure 7.122 *Rhyacophila nigrorosea* (♀): (A) Lateral view of female genitalia; (B) Dorsal view of segment VIII; (C) Ventral view of segment VIII; (D) Lateral view of vaginal apparatus; (E) Ventral view of vaginal apparatus.

Holotype & Allotype: PAKISTAN: Himalaya: Kawai, 24-VI-1953; Mahandri, 26-VI-1953; Dunga Nar, 27-VII-1953: Surgun, 29-30-VII-1953; Kel, 16-VIII-1953; Janwai, 29-VIII-1953; Balakot, 16-X-1953; Kawai, 17-X-1953; Kanur, 13-V-1954. Hindu-Kush: Khoghozi, 3-5-X-1951.

Holotype depository: Zoological Museum of Lausanne, Switzerland.

Additional material examined: INDIA: Himachal Pradesh: Kothi, 1900 m, 06-VII-2009, 1♀; Uttarakhand: Auli, 2900 m, 16-VI-2011, 1♀. (Coll. Parey & Pandhar).

Distribution: Pakistan, India (Uttarakhand, Himachal Pradesh).

109. *Rhyacophila norbu* Schmid

Rhyacophila norbu, Schmid, 1970: 147

Adult: Dorsal side of the body dark brown, with fine concolorous hairiness. Pleura and legs light red. Forewings tinder golden, very dark and riddled with silvery macules, clearly visible. Wingspan: 12.5–14.5 mm.

Genitalia (♂): (Figs. 7.123A–D). Dorsally, the Xth segment appears as a narrow oval, simple and slightly indented at its extremity; seen from the front, it shows three small spines of unequal lengths, directed downwards and slightly divergent. Second segment of the inferior appendages large, wide and slightly indented at its end. Dorsal branch of phallus well developed.

Genitalia (♀): Not available.

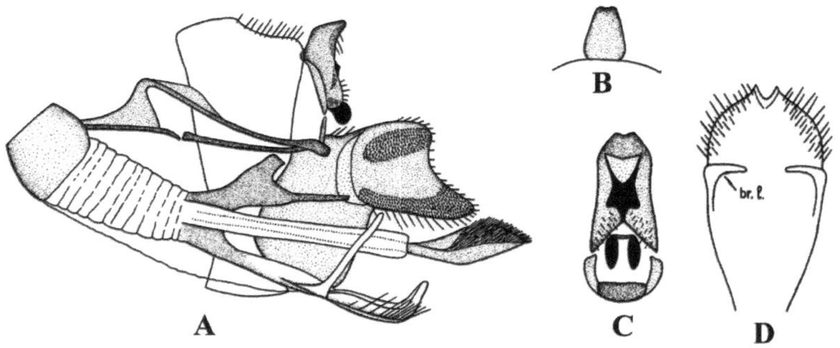

Figure 7.123 *Rhyacophila norbu* (♂): (A) Lateral view of male genitalia; (B) Dorsal view of Xth segment; (C) Front view of Xth segment; (D) Ventral plate of phallus, dorsal view (from Schmid, 1970).

Holotype: INDIA: Assam, United Jaintia and Khasi Hills, Demthring, 3000–3800 ft., 16-IV-1960.

Holotype depository: ROM, Canada.

Distribution: India (Meghalaya).

110. *Rhyacophila nyamangpa* Schmid

Rhyacophila nyamangpa, Schmid, 1970: 194

Adult: Upper body dark brown with colored hairs. Legs yellow-brown. Wings uniformly black-brown. Wingspan: 21–25 mm.

Genitalia (♂): (Figs. 7.124A–B). Preanal appendages in elongated ovals, vertically truncated at their extremity and with their upper apical angle a little projecting. Complex-shaped phallus, with its upper edge broadened horizontally over its middle third and provided with a lower median lobe; parameres in simple spines and curved at their base. Second segment of the inferior appendages terminated in two obtuse lobes of similar shapes, but of very different lengths.

Genitalia (♀): Not available.

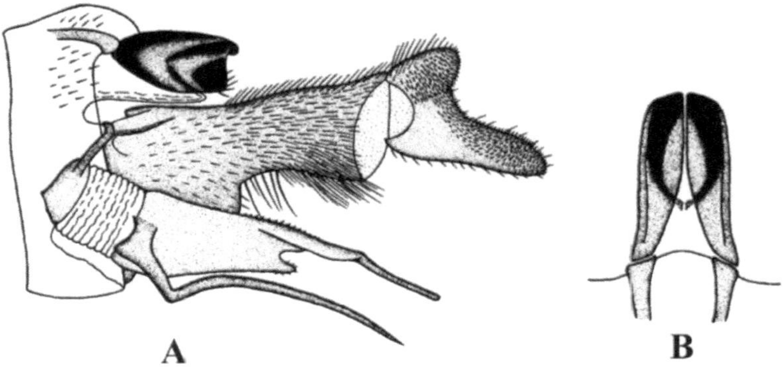

A B

Figure 7.124 *Rhyacophila nyamangpa* (♂): (A) Lateral view of male genitalia; (B) Dorsal view of Xth segment (from Schmid, 1970).

Holotype: INDIA: Sikkim, Karponang, 8000–10000 ft., 22-VIII-1959.

Holotype depository: ROM, Canada.

Distribution: India (Sikkim).

111. *Rhyacophila nyelungpa* Schmid

Rhyacophila nyelungpa, Schmid, 1970: 183

Adult: Top of the body reddish-brown, with concolorous hairiness. Light reddish-brown pleura and legs. Wings uniformly reddish-brown. Wingspan: 15–20 mm.

Genitalia (♂): (Figs. 7.125A–C). Apical dorsal lobe of segment IX slightly elongated. Preanal appendages sub-quadrangular, with an apical notch. Anal sclerites small but tall. Dorsal appendages of the phallic apparatus very slightly projecting; slender aedeagus (phallus), with a four-lobed ventral plate; spiniform and setiferous parameres.

Genitalia (♀): Not available.

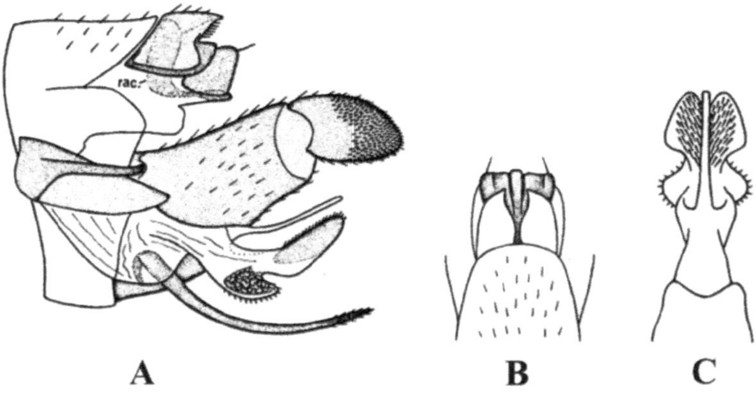

Figure 7.125 *Rhyacophila nyelungpa* (♂): (A) Lateral view of male genitalia; (B) Dorsal view; (C) Phallic apparatus, dorsal view (from Schmid, 1970).

Holotype: INDIA: Assam, United Jaintia and Khasi Hills, Syndai, 14-XII-1959.

Holotype depository: ROM, Canada.

Distribution: India (Meghalaya).

112. *Rhyacophila nyerongpa* Schmid

Rhyacophila nyerongpa, Schmid, 1970: 204

Adult: Top of the body light reddish-brown, with concolorous hairiness. Pleura and legs reddish yellow; gray tarsi. Forewings dark brown, with a few light spots, quite large, clearly visible, but with not very clear contours. Wingspan: 14–15 mm.

Genitalia (♂): (Figs. 7.126A–B). Apical dorsal part of the IXth segment with a sparse brush of hairs. Parameres extremely long and slender, not thickened at the end. Second segment of the inferior appendages with a very large circular notch whose edge is lined with a large area of black spinules.

Genitalia (♀): Not available.

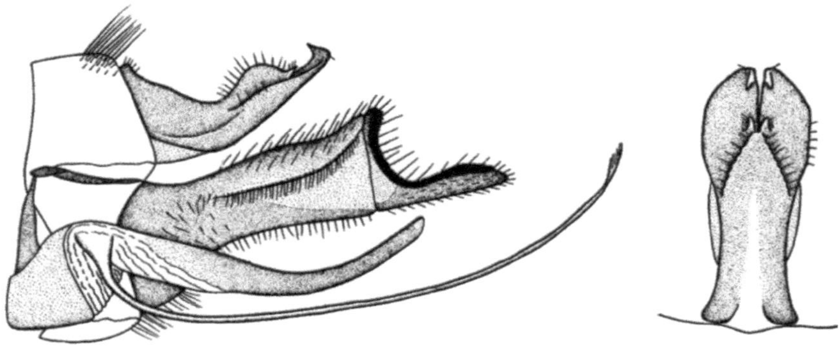

A B

Figure 7.126 *Rhyacophila nyerongpa* (♂): (A) Lateral view of male genitalia; (B) Dorsal view (from Schmid, 1970).

Holotype: INDIA: Assam: NEFA, Kameng Frontier Division; Arunachal Pradesh: Nakhu, 6-VII-1961.

Holotype depository: ROM, Canada.

Distribution: India (Assam, Arunachal Pradesh).

113. *Rhyacophila nyerpa* Schmid

Rhyacophila nyerpa, Schmid, 1970: 156

Adult: Dorsal side of the body yellow-brown. Wings black-brown. Wingspan: 16–22 mm.

Genitalia (♂): (Figs. 7.127A–C). Segment IX quite elongated and with a short apical dorsal lobe. Horizontal part of the Xth segment long; vertical part inserted at the base of the previous one. Phallus partially fused to its ventral plate, which is broad, strongly concave and not deeply indented at its extremity. Second segment of the inferior appendages triangular and well individualized.

Genitalia (♀): Not available.

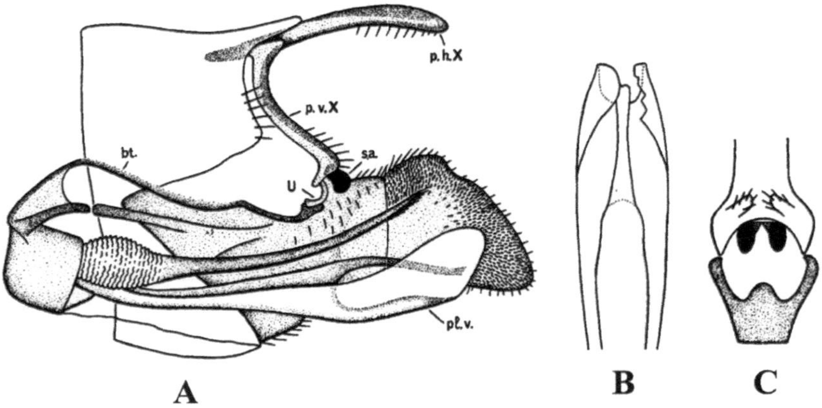

Figure 7.127 *Rhyacophila nyerpa* (♂): (A) Lateral view of male genitalia; (B) Dorsal view of phallic apparatus; (C) Front view of Xth segment (from Schmid, 1970).

Holotype: INDIA: Assam: NEFA, Kameng Frontier Division; Arunachal Pradesh: Jhum La, 6800–9000 ft., 24-25-III-1961.

Holotype depository: ROM, Canada.

Distribution: India (Assam, Arunachal Pradesh).

114. *Rhyacophila nyukmadongpa* Schmid

Rhyacophila nyukmadongpa, Schmid, 1970: 166

Adult: Dorsal side of the body and wings entirely and uniformly black, with colorless hairiness. Dark brown pleura and legs. Relatively broad wings. Wingspan: 16–19 mm.

Genitalia (♂): (Figs. 7.128A–B). Dorsal part narrowed, truncated at its extremity and armed with three strong teeth in lower subapical position. Anal sclerites sub-quadrangular and distinctly trilobed. Sagittal appendage very short and wide triangular. Second segment of the inferior appendages regular and obtuse half-oval.

Genitalia (♀): Not available.

Holotype: INDIA: Assam, NEFA, Kameng Frontier Division; Arunachal Pradesh: Nyukmadong, 6600–8000 ft., 23-IV-1961.

Holotype depository: ROM, Canada.

Distribution: India (Assam, Arunachal Pradesh).

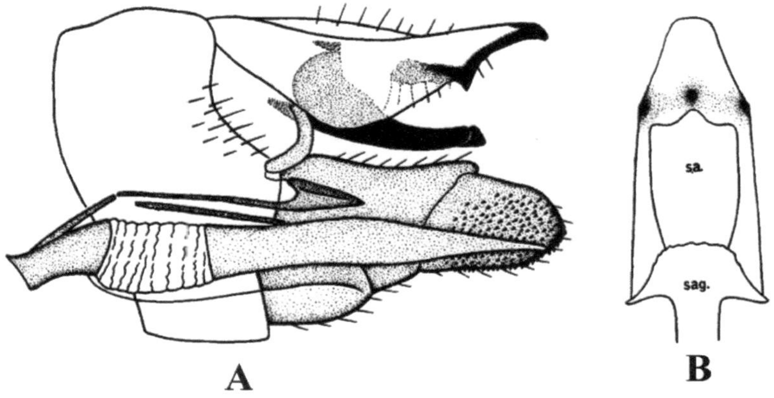

Figure 7.128 *Rhyacophila nyukmadongpa* (♂): (A) Lateral view of male genitalia; (B) *Dorsal view* (from Schmid, 1970).

115. *Rhyacophila obscura* Martynov

Rhyacophila obscura, Martynov, 1927: 163–164

Adult: Head, thorax and legs brown; rest of body pale brown and antenna with alternating light and dark brown bands. Length of antenna 4.3 mm and forewing 6.79 mm. In forewing fork 2nd arise earlier than 1st. In females, entire body brown except scape and pedicel of antenna which are pale brown and flagellum with alternating dark and light brown bands.

Genitalia (♂): (Figs. 7.129A–C). In dorsal view, segment X trilobed, outer lobes broader than median lobe. Outer lobes of segment X rounded whereas middle lobe is triangular shaped. Tergum of segment IX larger than its sternum. Segment X with rounded apex. Anal sclerite rounded. Phallicata with ventral lobe which is curved upwards. Paramere bulbous clothed with setae. Basal segment of inferior appendage longer than terminal one. Superior lobe of terminal segment of inferior appendage broader than inferior lobe. Superior lobe convexly raised and inferior lobe digitiform. In ventral view basal segment of inferior appendage with concavity.

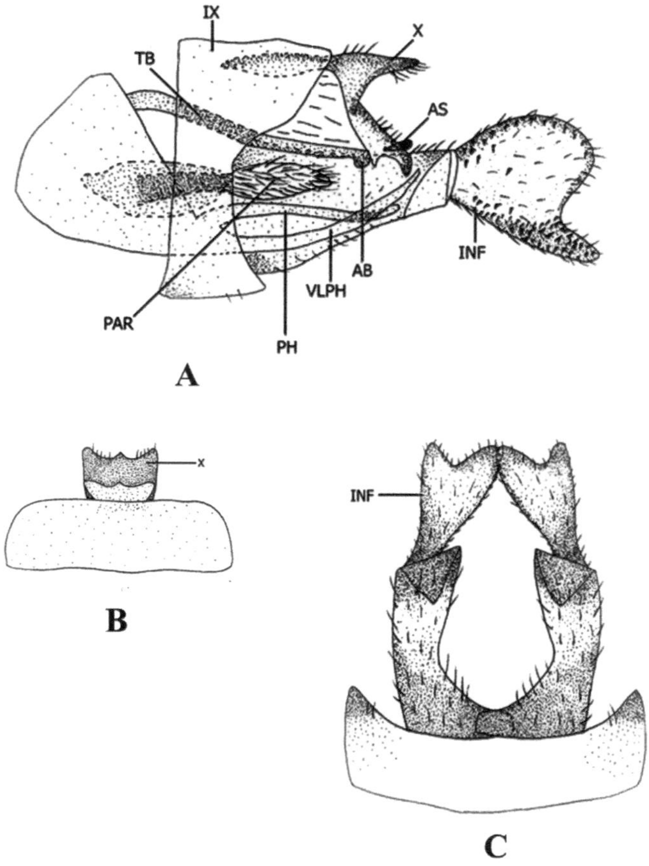

Figure 7.129 *Rhyacophila obscura* (♂): (A) Lateral view of male genitalia; (B) Dorsal view; (C) Ventral view.

Genitalia (♀): (Figs. 7.130A–C). VIth segment with spine ventrally and tergum and sternum of segment VII rounded. In lateral view, VIIIth segment long and tapering posteriorly with dorsal margin shorter than ventral. Pair of long slender apodemes extending anteriorly to middle of VIth segment. Posterodorsal margin of VIIIth segment with V-shaped indentation whereas its posteroventral margin deeply and narrowly notched. Ventrally, posterior process of vaginal apparatus triangular at the apex; intermediate process divided into lobes and longer than posterior process.

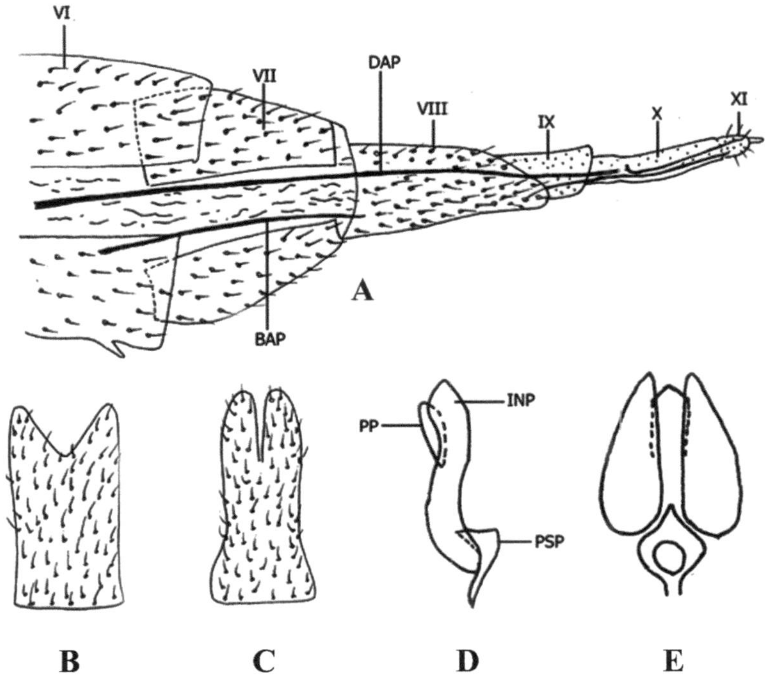

Figure 7.130 *Rhyacophila obscura* (♀): (A) Lateral view of female genitalia; (B) Dorsal view of segment VIII; (C) Ventral view of segment VIII; (D) Lateral view of vaginal apparatus; (E) Ventral view of vaginal apparatus.

Holotype: INDIA: Turkestan (Uzbekistan): Tchimgan, 14-28-VII-1924.

Holotype depository: Ann. Mus. Zool. Acad. Sci., USSR.

Additional material examined: INDIA: Jammu & Kashmir: Pahalgam, 2100 m, 28.viii.2008, 1♂; Aru, 2400 m, 29.viii.2008, 1♀; Gulmarg, 2900 m, 02.viii.2009, 2♂♂, 1♀; Dachigam, 1800 m, 07.viii.2009, 1♂; Varinag, 2300 m, 12.viii.2009, 1♂, 1♀; Pahalgam, 2100 m, 14.viii.2009, 2♂♂, 2♀♀; Pahalgam, 2100 m, 15.viii.2009, 1♀; Pahalgam, 2100 m, 12.viii.2010, 3♂♂, 2♀♀; Aru, 2400 m, 20.vii.2011, 5♂♂, 9♀♀; Sonmarg, 2700 m, 26.vii.2011, 3♂♂, 25♀♀; Tangmarg, 2600 m, 15.ix.2012, 1♀; Arunachal Pradesh: Lumla, 2300 m, 15.v.2011, 2♂♂, 1♀; Sikkim: Lachen, 3200 m, 24.v.2011, 6♂♂, 5♀♀; Himachal

Pradesh: Saho, 1900 m, 13.ix.2008, 1♂, 16♀♀; Gehra, 1200 m, 14.ix.2008, 2♂♂, 1♀; Holi, 1650 m, 16.ix.2008, 2♂♂; Baksunag, 2300 m, 25.v.2008, 6♂♂, 7♀♀; Keylong, 3100 m, 10.viii.2008, 2♂♂, 1♀; Kund, 1100 m, 11.vi.2009, 1♂; Ghiaghi, 2200 m, 29.vi.2009, 4♂♂, 7♀♀; Banjar, 950 m, 30.vi.2009, 3♂♂; Kasol, 1400 m, 1.vii.2009, 1 ♂; Manikaran, 1700 m, 02.vii.2009, 1♀; Barot, 2500 m, 06.vi.2010, 1♂, 2♀♀; Ahla, 2100 m, 11.vii.2010, 1♂, 2♀♀; Barara, 1800 m, 16.vii.2010, 7♂♂; Traila, 1800 m, 17.vii.2010, 6♂♂, 5♀♀; Satrundi, 2700 m, 15.vi.2011, 1♂, 3♀♀; Sanghani, 2100 m, 18.vi.2011, 11♂♂, 3♀♀; Traila, 1800 m, 19.vi.2011, 4♂♂, 3♀♀; Khajjiar, 1950 m, 13.viii.2011, 5♂♂ Uttarakhand: Mornala, 2200 m, 11.vi.2008, 1♀; Badrinath, 3400 m, 24.ix.2008, 1♂, 8♀♀, Shyna Chatti, 2000 m, 27.ix.2008, 2♂♂, 3♀♀; Maneri, 3300 m, 05.vi.2009, 1♂, 1♀; Tala, 1800 m, 12.vi.2009, 10♂♂, 2♀♀; Mandal, 1400 m, 14.vi.2009, 1♂, 3♀♀; Pothibasa, 2200 m, 15.vi.2010, 4♂♂; Mandal, 1400 m, 17.vi.2010, 2♂♂, 2♀♀; Chopta, 2700 m, 18.vi.2010, 2♀♀; Loharkhet, 1700 m, 19.vi.2011, 1♂; Mana, 3200 m, 23.vi.2011, 6♀♀; Badrinath, 3400 m, 22.vi.2011, 3♂♂, 7♀♀; Gobinddham, 3100 m, 18.vi.2012, 2♂♂. (Coll. Parey & Pandhar).

Distribution: Turkestan (Uzbekistan), Bhutan, Nepal, Afghanistan, Pakistan, India (Himachal Pradesh, Sikkim, Jammu and Kashmir, Arunachal Pradesh, Uttarakhand).

116. *Rhyacophila parva parva* Kimmins

Rhyacophila parva parva, Kimmins, 1953: 536

Subspecies: *Rhyacophila parva mukpo,* Schmid, 1970: 174

Adult: Body dark brown, with blackish hairs on vertex and thorax. Legs fuscous. Wings elongate, membrane brownish, with brownish pubescence.

Genitalia (♂): (Figs. 7.131A–B). IXth segment with an acute excision on each lateral margin at the base of the dorsal process, which is moderately long, from above with its lateral margins convex, tapering from beyond midway to a broad, slightly convex apex. Xth tergite from above consisting of two small, triangular lobes with an acute excision between them, the lower margins of the lobes fused to form a scoop. Basal sleeve of phallus produced apically in a pair of plates with dilated apices, each apex with a plate-like dilatation on its inner surface. Phallus with an upcurved spine at its apex; from the upper surface of the spine, near its base, arises a thin, upwardly directed plate. The lower surface of the phallus is developed into two curved claws, broad and foliate at their bases, lined internally with bristles, and with a basally directed, membranous extension. Inferior appendage with basal segment moderately long, deep at its base, tapering to about half its width at apex. Terminal segment short, subtriangular, with a rounded apex.

Genitalia (♀): Not available.

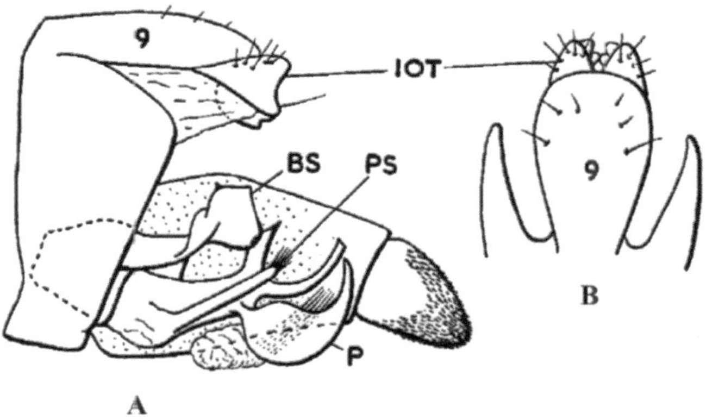

Figure 7.131 *Rhyacophila parva* (♂): (A) Male genitalia, lateral view; (B) Dorsal view of IXth and X[th] segments (from Kimmins, 1953).

Holotype: INDIA: Assam: Khasi Hills, Cherrapunji.

Holotype depository: BMNH, London.

Distribution: India (Meghalaya).

117. *Rhyacophila paurava* Schmid

Rhyacophila paurava, Schmid, 1959: 237

Adult: Head, top of the thorax, antennae and palpi black, with concolorous hairiness. Pleura and paws brown. Abdomen dark brown above and yellow-brown below. Wings entirely and uniformly black-brown, the hindwings being slightly less dark than the forewings.

Genitalia (♂): (Figs. 7.132A–B). Segment IX quite regularly elongated on all its circumference, except dorsally where it extends into a large well individualized plate and deeply bifid at the end. Xth segment forming semi-ovoid lateral lobes flanking the dorsal part of the IXth segment and a concave part in V. Basal segments of the inferior appendages 2.5 times longer than high and with the lower edge a little concave. Terminal segments in simple triangles, directed obliquely downwards, with the apical margin weakly domed and furnished with a narrow band of fine spinules. Penal apparatus composed of three parts; phallus in the shape of a small slender tube, slightly chitinous and in a ventral position; titillators in large ergots, curved upwards, blunt and strongly chitinous; there is also an odd dorsal appendage, slender, long, with irregular outlines and rather strongly hairy.

Genitalia (♀): Not available.

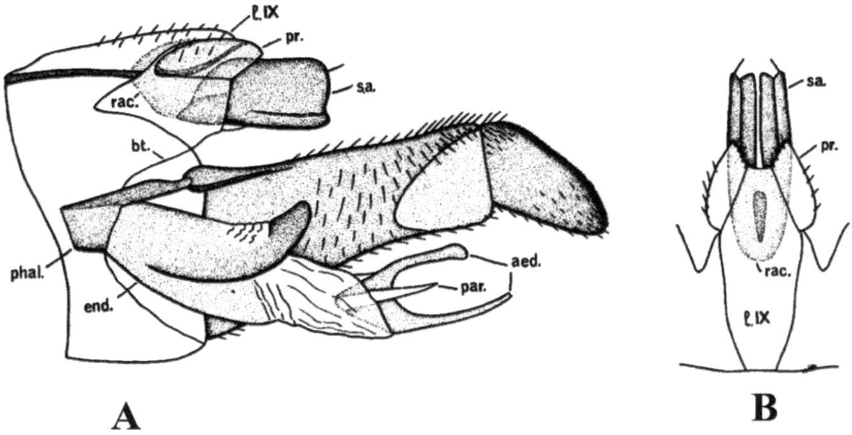

Figure 7.132 *Rhyacophila paurava* (♂): (A) Lateral view of male genitalia; (B) *Dorsal view* (from Schmid, 1970).

Holotype: PAKISTAN: Murree, 8-VI-1953; Mahandri, 26-VI-1953; Kaghan, 27-29-VI-1953; Naran, 30-VI, 5-VII-1953; Surgun, 29-30-VII-1953; Shardi, 1-13-VIII-1953.

Holotype depository: Zoological Museum of Lausanne, Switzerland.

Distribution: Pakistan, India (Meghalaya, Uttarakhand).

118. *Rhyacophila pemba* Schmid

Rhyacophila pemba, Schmid, 1970: 147–148

Adult: Dorsal side of the body dark brown, with concolorous hairiness. Pleura and legs red. Tinder forewings very dark and riddled with rather dense golden macules. Wingspan: 14–19 mm.

Genitalia (♂): (Figs. 7.133A–D). Xth segment appearing obtuse, seen in profile, but characteristic, seen from the back, by very chitinous ridges forming a design reminiscent of two reversed and parallel Ys. Second segment of the inferior appendages forming three slightly protruding lobes. Phallus with its dorsal branch reduced to a simple bulge.

Genitalia (♀): Not available.

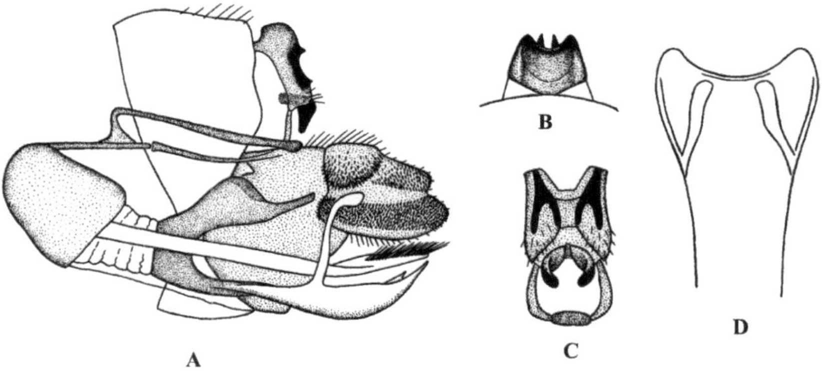

Figure 7.133 *Rhyacophila pemba* (♂): (A) Lateral view of male genitalia; (B) Dorsal view of Xth segment; (C) Front view of Xth segment; (D) Ventral plate of phallus, dorsal view (from Schmid, 1970).

Holotype: INDIA: Assam: United Jaintia and Khasi Hills, Tarangblang, 200–5000 ft., 27-XII-1959.

Holotype depository: ROM, Canada.

Distribution: India (Meghalaya).

119. *Rhyacophila poba* Schmid

Rhyacophila poba, Schmid, 1970: 218–219

Adult: Body entirely russet brown, paler below. Remarkably long and slender legs. Wings very large, long and narrow.

The forewings are yellow, finely and intensely riddled with small dark macules, regularly distributed, but there are dark and light areas, larger in the center of the wing and in the post-costal area.

Genitalia (♂): (Figs. 7.134A–B). Apical dorsal lobe of segment IX broad, but thin. Segment X appearing quadrangular in lateral view, with its superior apical angle prominent. Second segment of the inferior appendages remarkable for the stretching of the lower lobe; the apical indentation is at a right angle and its edges are lined with a wide band of tiny red tubercles, of a velvety appearance.

Genitalia (♀): Not available.

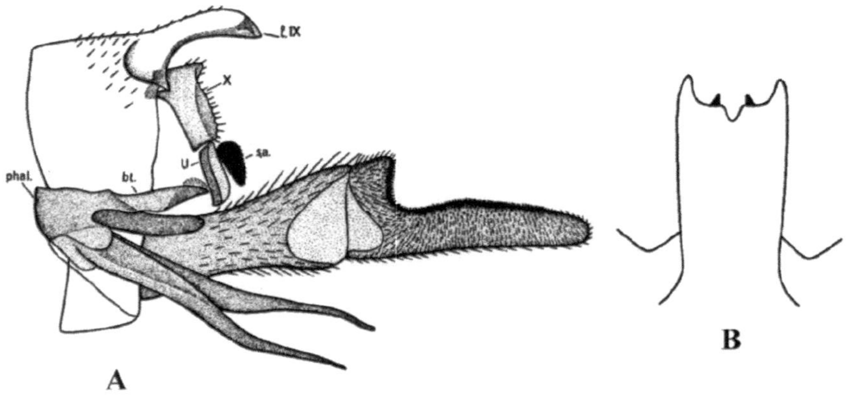

Figure 7.134 *Rhyacophila poba* (♂): (A) Lateral view of male genitalia; (B) Apical dorsal lobe of IXth segment (from Schmid, 1970).

Holotype: INDIA: Sikkim: Tangshing, 12000–13000 ft., 15-IV-1959.

Holotype depository: ROM, Canada.

Distribution: Tibet, India (Sikkim).

120. *Rhyacophila polha* Schmid

Rhyacophila polha, Schmid, 1970: 166

Adult: Above of the body dark brown, with concolorous hairiness. Dark reddish-brown pleura and legs. Uniformly black wings. Wingspan: 16–21 mm.

Genitalia (♂): (Figs. 7.135A–B). Wide dorsal piece, simple in shape and truncated straight at its extremity. Robust, elongated, bilobed anal scleritis. Elongated sagittal appendage. Upper apical angle of the first segment of the inferior appendages doubling the part of the second segment which is triangular and very weakly incised.

Genitalia (♀): Not available.

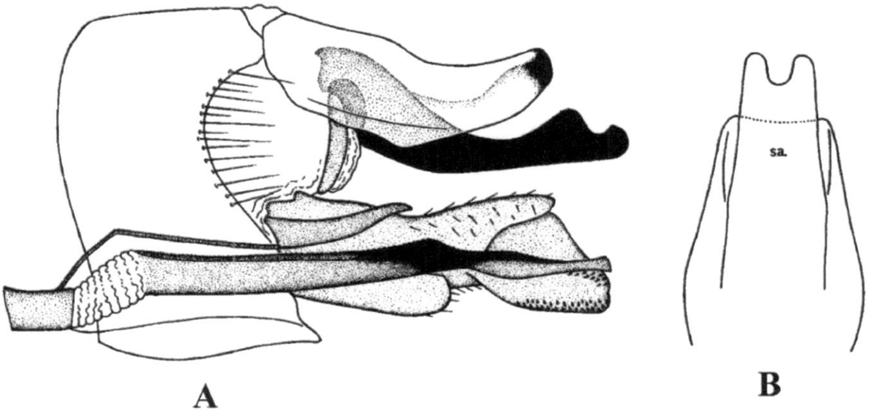

Figure 7.135 *Rhyacophila polha* (♂): (A) Lateral view of male genitalia; (B) *Dorsal view* (from Schmid, 1970).

Holotype: INDIA: Assam: NEFA, Kameng Frontier Division; Arunachal Pradesh: Domkho, 6900 ft., 11-V-1961.

Holotype depository: ROM, Canada.

Distribution: India (Assam, Sikkim).

121. *Rhyacophila procliva* Kimmins

Rhyacophila procliva, Kimmins, 1953: 542–543

Adult: Color of the body brownish, with brown and golden hairs. Legs brownish, anterior and median tibiae dark brown towards the apices. Wings elongate, rather pointed; anterior brownish, pubescence dark brown and golden, membrane with numerous hyaline irrorations. Posterior wing smoky hyaline, with brown pubescence. Abdomen yellowish-brown beneath, reddish-purple above, genitalia yellowish-brown.

Genitalia (♂): (Figs. 7.136A–C). IXth segment much produced in a long, narrow, hooded dorsal process, slightly constricted basally and with a truncate apex. Xth tergite short, situated beneath the dorsal process, forming a downwardly sloping hood. At its lower angles are two very small fingers with setae at their apices, and hinged beneath them is a small, deeply bifid scoop. Xth sternite comprises a narrow, sclerotized half-ring, attached to a slender band arising from the base of the phallus. Phallus long, slender, particularly beyond midway, and with a sinuous bend at the apex. Sheaths long, spiniform, with bifid apices. Inferior appendage long and slender. Basal segment broad at its base, tapering to midway and then parallel-sided, terminal segment short, deeply bifid, the arms divergent and unequal, the upper narrower and about half as long as lower, margins spinose.

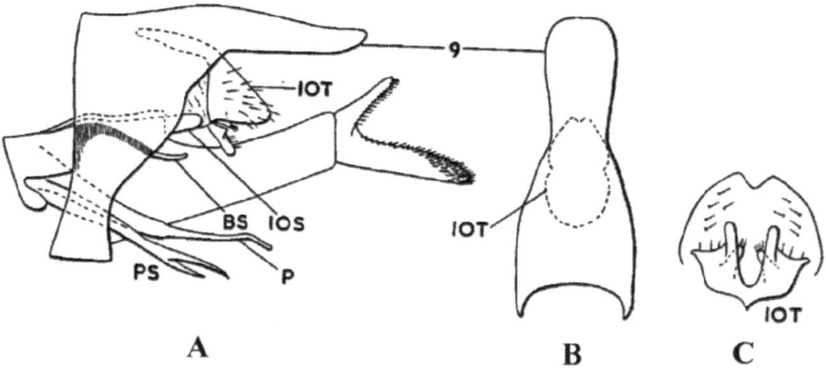

Figure 7.136 *Rhyacophila procliva* (♂): (A) Lateral view of male genitalia; (B) Dorsal view of IXth and Xth segments; (C) Ventral view of Xth segment (from Kimmins, 1953).

Genitalia (♀): (Figs. 7.137A–B). VIth sternite with a pointed ventral process. VIIIth segment rather longer than VIIth, from side with upper margin sloping downwards almost from base to an acute apex. Dorsal margin excised almost to base, ventral margin trilobed, outer lobes slightly longer than inner, incurved.

Holotype: MYANMAR: Kambaiti, 2000 m, 12-VII-1934.

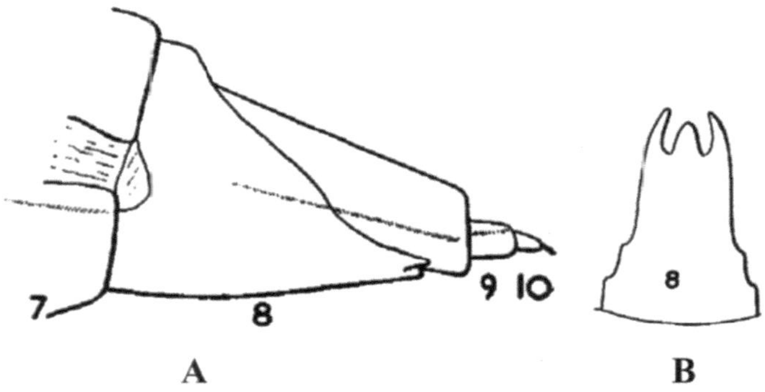

Figure 7.137 *Rhyacophila procliva* (♀): (A) Lateral view of VII–Xth sternites; (B) Dorsal view of IXth and Xth segments; (C) Ventral view of VIIIth sternite (Kimmins, 1953).

Holotype depository: BMNH, London.

Distribution: Myanmar, India (Manipur, Arunachal Pradesh).

122. *Rhyacophila putata putata* Kimmins

Rhyacophila putata, Kimmins, 1953: 512–513

Subspecies: *Rhyacophila putata temba,* Schmid, 1970: 167

Adult: Head and thorax dark with blackish hairs. Antennae fuscous, faintly annulated towards the base, with dense, short, golden-brown pubescence. Legs dark fulvous. Wings elongate, dark brown, densely pubescent. In the anterior, the membrane of the costal, subcostal and radial areas from base to level of stigma is paler. Abdomen brownish, anal appendages yellowish-brown.

Genitalia (♂): (Figs. 7.138A–C). IXth segment produced slightly at its ventral margin. Xth tergite large, elongate and hooded, from the side with a small shoulder near upper basal margin, deep at the base, tapering upwards to a short, hooked apex. Dorsally, the lateral margins are slightly concave, apex rounded, with a large, U-shaped, median excision. Xth sternite membranous. Phallus long, stout for most of its length, tapering near apex to a slender spine. No apparent sheaths. Inferior appendages shorter than Xth tergite, basal segments fused basally along their inner margins, short and quadrate from the side.

Genitalia (♀): Not available.

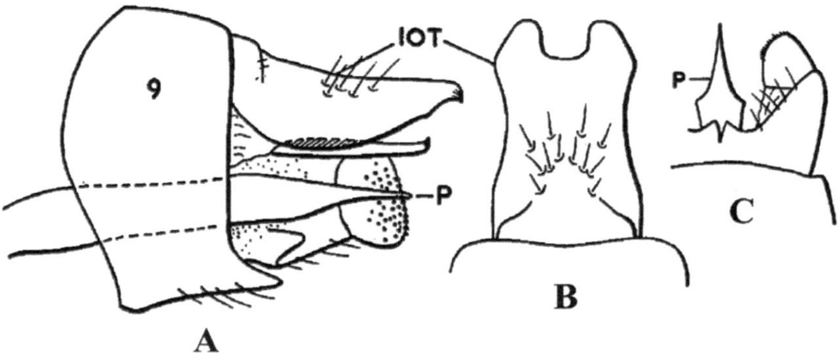

Figure 7.138 *Rhyacophila putata* (♂): (A) Lateral view of male genitalia; (B) Dorsal view of Xth tergite; (C) Ventral view of inferior appendage and phallus (from Kimmins, 1953).

Holotype: INDIA: Assam: Shillong, 5000 ft, 13-V-1924.

Holotype depository: BMNH, London.

Distribution: India (Assam, Meghalaya).

123. *Rhyacophila rhombica* Martynov

Rhyacophila rhombica, Martynov, 1935: 98–99

Adult: Yellowish-brown head; thorax dark brown dorsally, brown ventrally; legs yellowish, coxae brownish, tarsi densely clothed with short yellow hairs; anterior legs somewhat darker, brownish yellow. Anterior wings moderately narrow, brownish, with dark pterostigma; 2nd apical fork arises a little earlier than 1st. Posterior wings pale brownish; 2nd apical fork begin a little before 1st. Abdomen dark brown above, paler beneath.

Genitalia (♂): (Figs. 7.139A–B). IXth tergite separated by two very deep triangular excisions but containing pale membranes. In dorsal view, IXth tergite is subquadrate, with two brown narrow stripes which diverge from the base to the posterior outer angles. Basal segment of inferior appendage elongated, broad from side, narrower from above, with raised straight upper edge; second segment triangular from side, with extended hind-lower angle; hind margin clothed with dense brush of short hairs; upper basal portion of 2nd joint curved inwards; its inner margin, as well as the whole hind margin of the joint is dorsally beset with a series of short spinules. On the sides of phallus are two short, but thick appendages, curved upwards and black at their ends.

This species is evidently allied to *Rhyacophila choprai* due to the similarity in the 9th segment (Martynov, 1935).

Genitalia (♀): Not available.

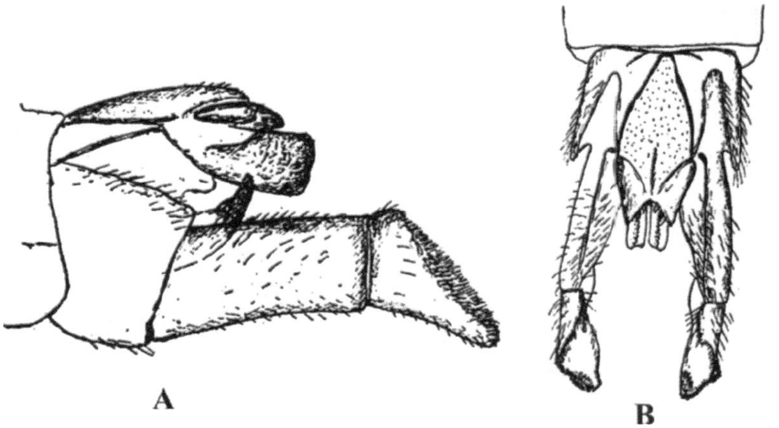

Figure 7.139 *Rhyacophila rhombica* (♂): (A) Lateral view of male genitalia; (B) Dorsal view (from Martynov, 1935).

Holotype: INDIA: West Bengal: Darjeeling, 7000 ft., 7-VI-1917.

Distribution: Bhutan, Vietnam, Nepal, India (West Bengal, Himachal Pradesh).

124. *Rhyacophila rongpa* Schmid

Rhyacophila rongpa, Schmid, 1970: 222–223

Adult: Entire body pale brown except metathorax and abdomen; Fork 1st of forewing with dark brown patch. In forewing fork 2nd arise earlier than 1st.

Genitalia (♂): (Figs. 7.140A–C). In dorsal view, apices of dorsal apical lobe of segment IX curved inwards. In lateral view, tergum of segment IX longer than sternum. Dorsal apical lobe of segment IX hook shaped and curved downwards. Segment X lying beneath dorsal apical lobe of segment IX. Anal sclerite elongated and slightly hidden below segment X.

Phallicata rounded apically and paramere curved downwards. Inferior appendage three times longer than the segment IX and X together. Basal segment of inferior appendage broader at base. Terminal segment of inferior appendage with superior lobe oblique and inferior lobe two times longer than the superior lobe. In ventral view, inferior appendages slender.

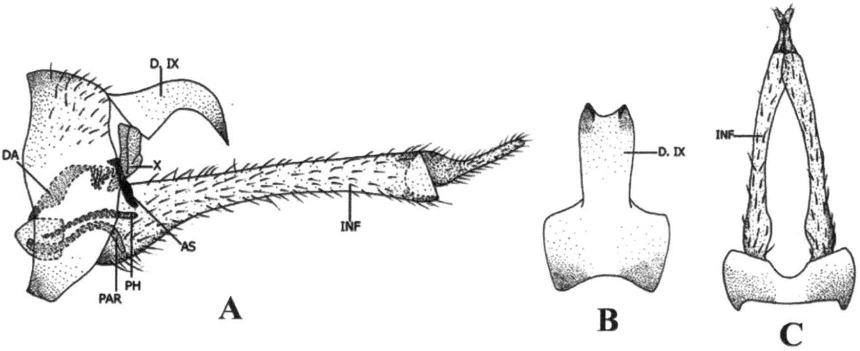

Figure 7.140 *Rhyacophila rongpa* (♂): (A) Lateral view of male genitalia; (B) Dorsal view; (C) Ventral view.

Genitalia (♀): (Figs. 7.141A–E). In lateral view, tergum as well as sternum of segment VIII same in length. Posterolaterally, segment VIII truncate. In dorsal view, segment VIII with indentation and in ventral view, segment VIII truncate. Intermediate process of vaginal apparatus longer than all other parts of vaginal apparatus. In ventral view, processus spermathecae leaf shaped.

Holotype: INDIA: Manipur: Sirohi Kashong, 6-7-VI-1960.

Holotype depository: ROM, Canada.

Additional material examined: INDIA: Arunachal Pradesh: Dirang, 1700 m, 06-X-2010; Hunli, 1450 m, 02-V-2011; Uttarakhand: Badrinath, 3400 m, 22-VI-2011. (Coll. Parey & Pandhar).

Distribution: Bhutan, Nepal, India (Manipur, Arunachal Pradesh, Uttarakhand).

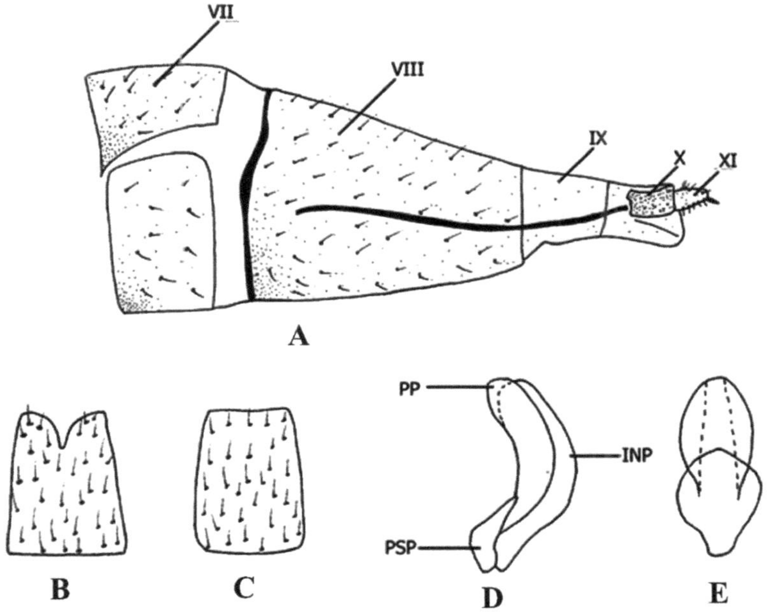

Figure 7.141 *Rhyacophila rongpa* (♀): (A) Lateral view of female genitalia; (B) Dorsal view of segment VIII; (C) Ventral view of segment VIII; (D) Lateral view of vaginal apparatus; (E) Ventral view of vaginal apparatus.

125. *Rhyacophila sajadi* Saini and Kaur

Rhyacophila sajadi, Saini and Kaur, 2012: 309–310

Adult: Head, antenna, maxillary palp and labial palp pale brown with black pubescence. Prothorax and metathorax pale brown but mesothorax dark brown. Abdomen and legs dark brown in color. Length of antenna 10.4 mm and forewing 14.0 mm and pale brown with brown irrotations but hindwing light brown. In forewing forks 1st and 2nd arise at the same level whereas in females, fork 2nd arises slightly earlier than 1st.

Genitalia (♂): (Figs. 7.142A–C). In dorsal view, apices of lateral processes of dorsal apical lobe of segment IX truncate; in lateral view, median process of dorsal apical lobe of segment IX pointed and lateral processes two times broader than median process apically and with medially directed triangular membranous projection in middle of its proximal half. Anal sclerite C-shaped. Dorsal appendage of phallic apparatus rounded at apex; phallicata curved downward. Dorsal surface of terminal segment of inferior appendage sloped; paramere absent. In ventral view, basal segment of inferior appendage with concavity at middle.

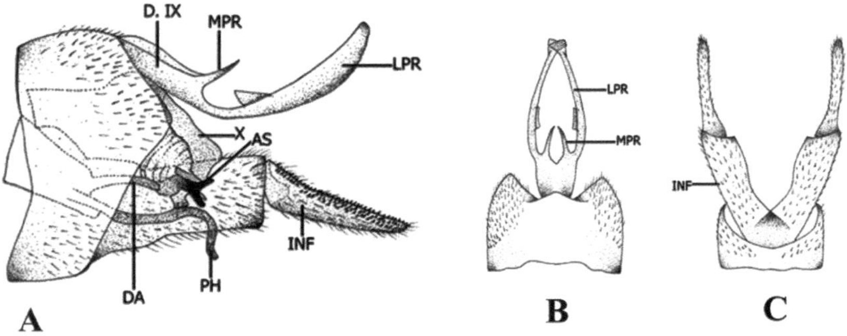

Figure 7.142 *Rhyacophila sajadi* (♂): (A) Lateral view of male genitalia; (B) Dorsal view; (C) Ventral view.

Genitalia (♀): (Figs. 7.143A–E). Simple, slender, long and with conspicuous spine on sternum VI. Segment VII with distinct, setose tergum and sternum. VIIIth segment setose, roughly rectangular in shape; in lateral view, a pair of long slender apodemes extending anterad to segment VI. Intersegmental membranes between segments VII–VIII & VIII–IX are for protraction and retraction. Segment X and XI also membranous with terminal papilla; internally segment XI with slightly Y-shaped furcation of apodeme apically. Segment X ventrally with small sclerotized lobes. Segment VIII dorsally with slight indentation and ventrally sub-truncate in outline. In lateral view, vaginal apparatus with long posterior process with breadth of posterior process 1/3 that of processus spermathecae.

Holotype: INDIA: Uttarakhand: Badrinath, 3400 m, 22-VI-2011. (Coll. Parey & Pandhar).

Paratypes: INDIA: Uttarakhand: Badrinath, 3400 m, 22-VI-2011, 2♀♀; Munsiyari, 2400 m, 17-VI-2008, 1♀. Arunachal Pradesh: Dirang, 1700 m, 6-X-2010, 1♀. Sikkim: Sangkalang, 1750 m, 13-V-2009, 3♀♀. Assam: Jatinga, 800 m, 20-V-2010,

1♀. Himachal Pradesh: Satrundi, 2700 m, 15-VI-2011, 3♀♀. (Coll. Parey & Pandhar).

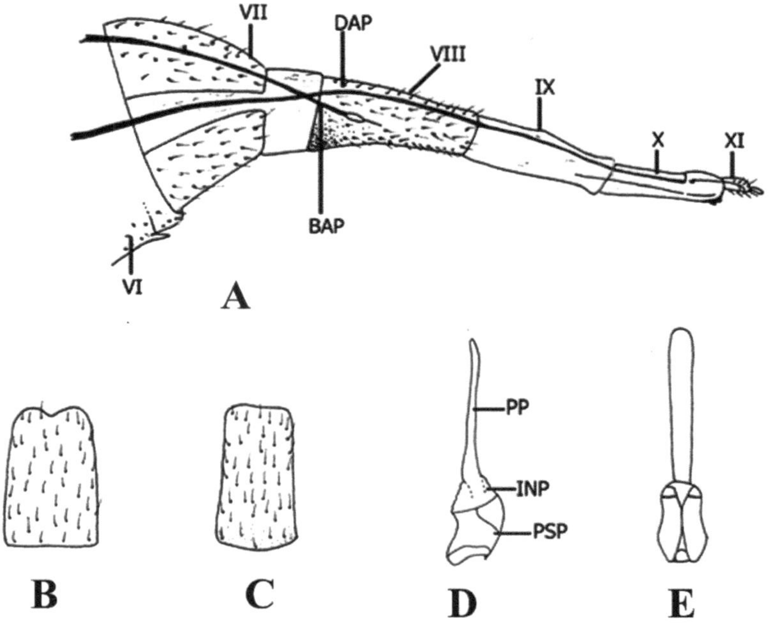

Figure 7.143 *Rhyacophila sajadi* (♀): (A) Lateral view of female genitalia; (B) Dorsal view of segment VIII; (C) Ventral view of segment VIII; (D) Lateral view of vaginal apparatus; (E) Ventral view of vaginal apparatus.

Holotype depository: MDZPUP, India.

Distribution: India (Uttarakhand, Arunachal Pradesh, Sikkim, Assam, Himachal Pradesh).

126. *Rhyacophila sakyapa* Schmid

Rhyacophila sakyapa, Schmid, 1970: 153

Adult: Entire body brown except labial palps which are pale brown. Length of antenna 7.76 mm and forewing 9.2 mm. In forewing forks 1st and 2nd arise at the same level.

Genitalia (♂): (Figs. 7.144A–C). In dorsal view segment X with U shape indentation. In lateral view, tergum IX with slight concavity near posterior margin and sternum is half in length as compared to tergum IX. Anal sclerites broad at apex. Parameres bulbous; phallicata smaller than the ventral lobe of phallicata. Apex of basal segment of inferior appendage 1/3rd in breadth as compared to base. Terminal segment of inferior appendage longer than the basal segment of inferior appendage. Superior side of terminal segment of inferior appendage convex and superior lobe broader than the inferior lobe. Inferior lobe slightly longer than superior lobe and both superior and inferior lobes with minute spinules. In ventral view, basal segment of inferior appendage with deep concavity near base.

Genitalia (♀): Not available.

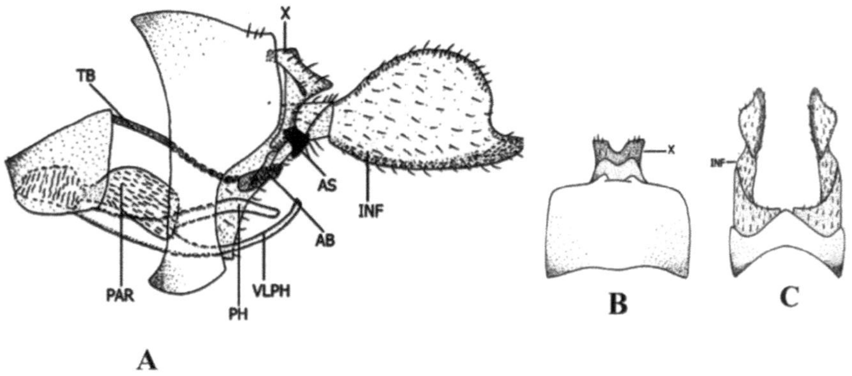

Figure 7.144 *Rhyacophila sakyapa* (♂): (A) Lateral view of male genitalia; (B) Dorsal view; (C) Ventral view.

Holotype: INDIA: Uttarakhand: Pauri Garhwal, Tarsali, 6-7-V-1958.

Holotype depository: ROM, Canada.

Additional material examined: INDIA: Uttarakhand: Sonprayag, 1500 m, 09-VI-2009; Pothibasa, 2200 m, 15-VI-2010. (Coll. Parey & Pandhar).

Distribution: Nepal, India (Uttarakhand).

127. *Rhyacophila sanglungpa* Schmid

Rhyacophila sanglungpa, Schmid, 1970: 202

Adult: Dorsal side of the body light reddish-brown, with golden hairs. Light yellow pleura and legs. Wings light orange-yellow, very few spotted with brown. Wingspan ♂ ♀: 17–18 mm.

Genitalia (♂): (Figs. 7.145A–C). Dorsal part of the Xth segment elongated and slender in shape. Dorsal appendage of phallic apparatus appearing oval when seen from above and narrowly cleft; parameres in slender rods and bearing four subapical claws. Second segment of the inferior appendages with its apical edge not notched, but sinuous.

Genitalia (♀): Not available.

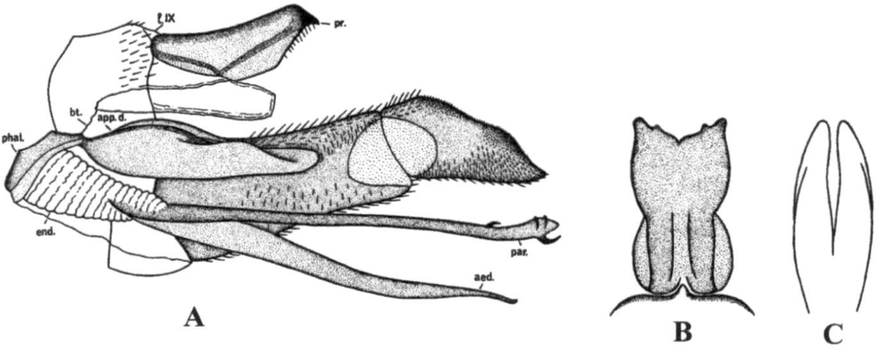

Figure 7.145 *Rhyacophila sanglungpa* (♂): (A) Lateral view of male genitalia; (B) Dorsal view; (C) Dorsal appendages of phallic apparatus (from Schmid, 1970).

Holotype: INDIA: Sikkim; Nanga, 3700–6500 ft., 3-4-VIII-1959.

Holotype depository: ROM, Canada.

Distribution: India (Sikkim).

128. *Rhyacophila scissa* Morton

Rhyacophila scissa, Morton, 1900: 5–6

Adult: Head, antenna, maxillary palp, thorax and abdomen brown but labial palp, legs and wings pale brown and forewing with brown irrotations. Length of antenna 8.73 mm and forewing 10.67 mm. In forewing fork 2nd arise earlier than 1st. In **females**, Entire body and wings dark brown and antenna with alternate dark and pale brown bands; prothorax pale brown. Length of antenna 9.0 mm and forewing 11.2 mm. In forewing fork 1st arise earlier than 2nd.

Genitalia (♂): (Figs. 7.146A–C). In dorsal view segment X with U shape indentation. In lateral view segment IX with little protruding dorsal apical lobe of segment IX. Segment X quadrangular in shape. Apex of anal sclerite rounded. Paramere rectangular in shape with tuft of setae apically. Phallicata bilobed and upper lobe smaller than the lower lobe. Terminal segment of inferior appendage with superior lobe finger like. Tergum of segment IX broader than its sternum. In ventral view apex of terminal segment of inferior appendage rounded.

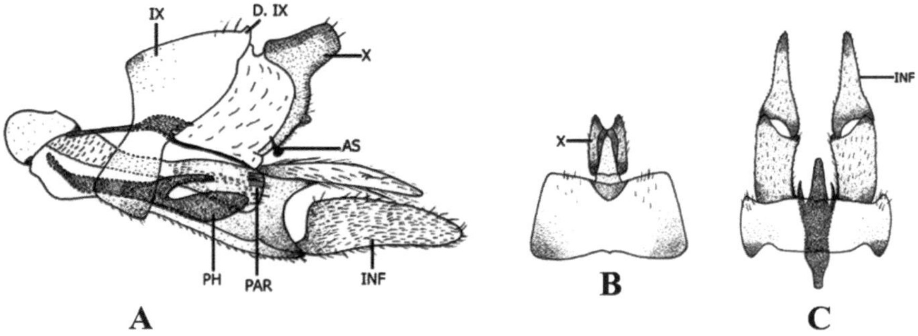

Figure 7.146 *Rhyacophila scissa* (♂): (A) Lateral view of male genitalia; (B) Dorsal view; (C) Ventral view.

Genitalia (♀): (Figs. 7.147A–E). Segment VI with spine ventrally. Dorsum of segment VII with long setae and length of dorsum as well as sternum is almost equal. Segment VIII with long setae on sternum. Dorsal apodeme extending anterad beyond segment VI and basal apodeme up to segment VII and it is bifurcated posteriorly. Segment IX, X and XI membranous. Tergum of segment VIII in dorsal view, deeply indented and in ventral view, it is with V shape indentation clothed with long setae and lobes are pointed in outline. In lateral view, posterior process of vaginal apparatus rounded, intermediate process broader apically and basally narrower.

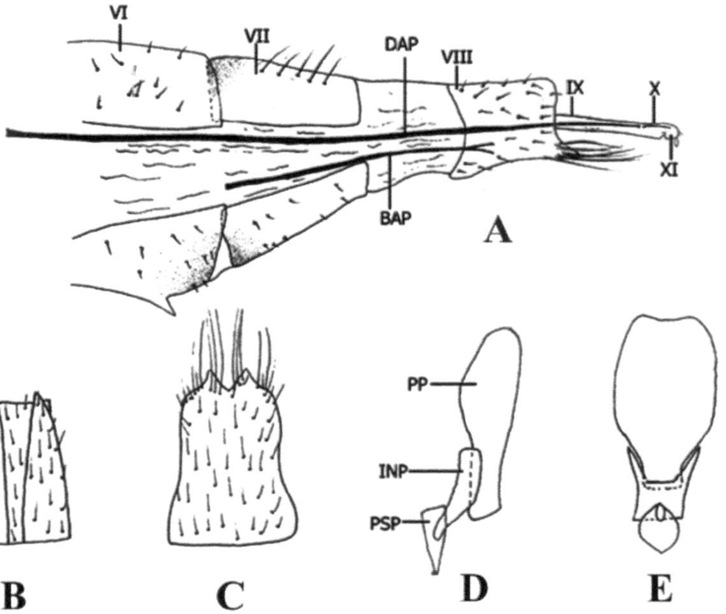

Figure 7.147 *Rhyacophila scissa* (♀): (A) Lateral view of female genitalia; (B) Dorsal view of segment VIII; (C) Ventral view of segment VIII; (D) Lateral view of vaginal apparatus; (E) Ventral view of vaginal apparatus.

Holotype: INDIA: Meghalaya; Khasia Hills, 6-XII-1899.

Holotype depository: BMNH, London.

Additional material examined: INDIA: Jammu & Kashmir: Pahalgam, 2900 m, 14.viii.2009, 2♂♂, 3♀♀; Arunachal Pradesh: Pene, 1200 m, 30.iv.2010, 2♂♂, 1♀; Dembuen, 2200 m, 05.v.2011, 1♂; Sikkim: Dentam, 1500 m, 10.v.2009, 3♂♂, 4♀♀; Dentam, 1500 m, 14.v.2011, 1♂; Nagaland: Pfutsero, 1900 m, 08.v.2010, 2♂♂; Himachal Pradesh: Satrundi, 2700 m, 15.vi.2011, 2♂♂, 1♀; Uttarakhand: Maneri, 1200 m, 05.vi.2009, 2♂♂, 4♀♀; Badrinath, 3400 m, 22.vi.2011, 1♂; Sagar, 1400 m, 14.vi.2011, 6♀♀. (Coll. Parey & Pandhar).

Distribution: Bhutan, Myanmar, Vietnam, China, Pakistan/ India (Meghalaya, Arunachal Pradesh, Sikkim, Uttarakhand, Jammu and Kashmir, Nagaland, Himachal Pradesh).

129. *Rhyacophila scissoides* Kimmins

Rhyacophila scissoides, Kimmins, 1953: 520

Adult: Head, thorax, wings and abdomen brown; antenna, maxillary palp and labial palp pale brown; legs orangish brown and forewing with dark brown irrotations. Length of antenna 4.85 mm and forewing 6.5 mm. In forewing fork 1st arise earlier than 2nd.

Genitalia (♂): (Figs. 7.148A–C). In dorsal view segment X with indentation. In lateral view tergum of segment IX longer than sternum. Segment X raised upward. Anal sclerite small and rounded apically. Phallicata trilobed, median lobe rounded, upper and lower lobes pointed. Superior lobe of inferior appendage finger like and inferior lobe rounded apically. Paramere extended, membranous and with tuft of setae apically. In ventral view basal segment of inferior appendage broader than terminal segment and the terminal segment rounded apically.

Genitalia (♀): Not available.

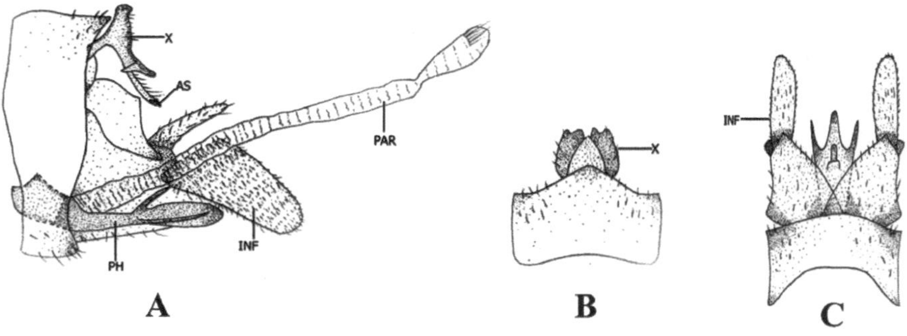

Figure 7.148 *Rhyacophila scissoides* (♂): (A) Lateral view of male genitalia; (B) Dorsal view; (C) Ventral view.

Holotype: INDIA: Assam; Khasia Hills, 1934.

Holotype depository: BMNH, London.

Additional material examined: INDIA: Arunachal Pradesh: Hunli, 1450 m, 02-V-2011, 3♂♂. (Coll. Parey & Pandhar).

Distribution: Vietnam, Nepal, China, Thailand, India (Meghalaya, Arunachal Pradesh, Sikkim, Uttarakhand).

130. *Rhyacophila scotina* Kimmins

Rhyacophila scotina, Kimmins, 1953: 513–514

Adult: Upper body dark red, with black hairs. Antennae dark fuscous, annulated with fulvous at the joints. Palpi and legs fuscous. Wings fuscous, posterior paler, pubescence fuscous. Apex of anterior wing elliptical, not pointed.

Genitalia (♂): (Figs. 7.149A–B). IXth segment with lateral margins sinuous, dorsal margin not produced. Xth tergite forming a long, narrow hood, tapering slightly towards a bluntly rounded apex. In lateral view, it is triangular, the lower margin on each side towards the apex with a blackened tooth. Ventrally, the Xth tergite is hinged a narrow, scoop-like plate, extending nearly to the apex of the Xth tergite. Xth sternite short, transverse and plate-like. Phallus simple, blade-like, with a narrowed apex. No apparent sheaths. Inferior appendages short, basal segments fused along their inner margins for about halfway. Terminal segment about half as long as basal, rounded from the side.

Genitalia (♀): Not available.

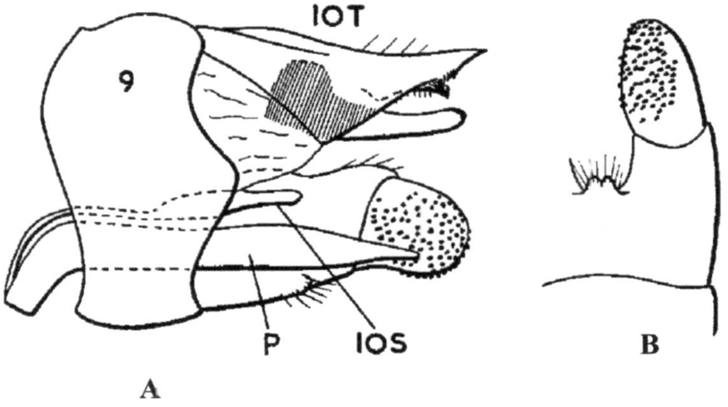

Figure 7.149 *Rhyacophila scotina* (♂): (A) Lateral view of male genitalia; (B) Inferior appendage, ventral view (from Kimmins, 1953)

Holotype: INDIA: Assam; Cherrapunji, 1934.

Holotype depository: BMNH, London.

Distribution: India (Meghalaya).

131. *Rhyacophila senggepa* Schmid

Rhyacophila senggepa, Schmid, 1970: 167

Adult: Small black insects, with pleura and legs brown. Wingspan: 23 mm.

Genitalia (♂): (Figs. 7.150A–B). Laterally, the dorsal part appears gradually thinned, broadly curved upwards, then downwards at its extremity; in dorsal view, it is shown to be a little narrowed at the level of its apical third and clearly incised at the apex. Long and slender anal scleritis, also a little indented. Second segment of the inferior appendages trapezoidal, higher than long and with its oblique lower edge.

Genitalia (♀): Not available.

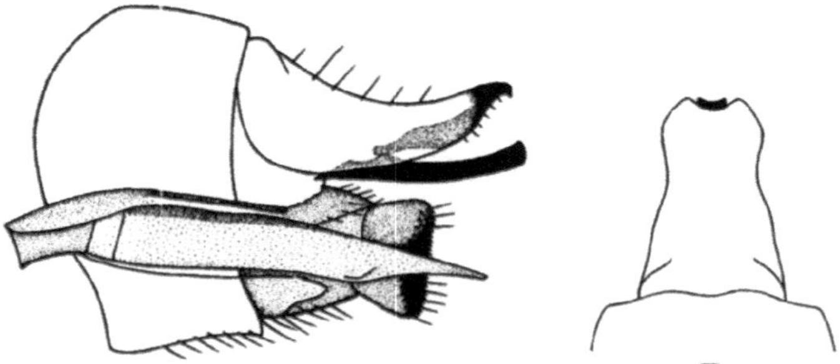

A B

Figure 7.150 *Rhyacophila senggepa* (♂): (A) Lateral view of male genitalia; (B) Dorsal view (from Schmid, 1970).

Holotype: INDIA: Assam; United Khasi and Jaintia Hills, Laitlyngkot, 17-III-1960.

Holotype depository: ROM, Canada.

Distribution: India (Meghalaya).

132. *Rhyacophila shakangpa* Schmid

Rhyacophila shakangpa, Schmid, 1970: 198–199

Adult: Body and legs light yellow, with concolorous hairiness. Light yellow-brown forewings, with rather large and inconspicuous golden spots. Wingspan: 13–19 mm.

Genitalia (♂): (Figs. 7.151A–B). Preanal appendages simple in form and entirely united by a membrane. Very short phallotheca; parameres in strong downward curved and spiniferous at their extremity; phallus in single cylinder invaginable in the base of the parameres. Second segment of the lower appendages high and short.

Genitalia (♀): Not available.

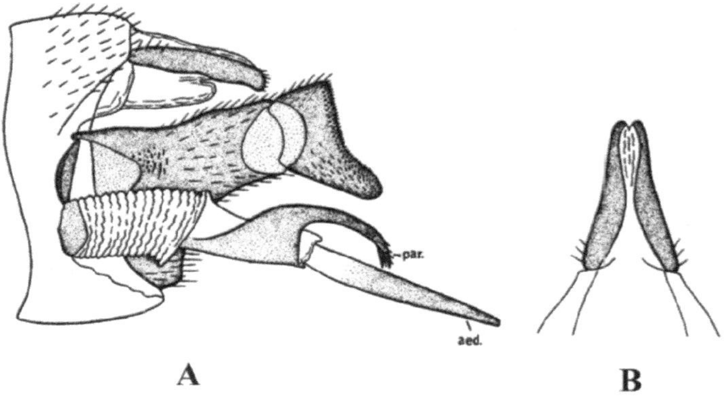

Figure 7.151 *Rhyacophila shakangpa* (♂): (A) Lateral view of male genitalia; (B) Preanal appendage, *dorsal view* (from Schmid, 1970).

Holotype: INDIA: Uttarakhand; Almora, Chhana, 22-IX-1958.

Holotype depository: ROM, Canada.

Distribution: Nepal, India (Uttarakhand).

133. *Rhyacophila sherchokpa* Schmid

Rhyacophila sherchokpa, Schmid, 1970: 208

Adult: Top of head greenish golden; golden rufous above thorax; hairs of the body light golden. Pleura and legs reddish yellow. Light golden forewings, dotted with dark brown macules, forming irregular patterns. Wingspan: 13–22 mm.

Genitalia (♂): (Figs. 7.152A–C). Apical dorsal part of the IXth segment forming two large lobes overhanging the base of the inferior appendages and each terminating in a small point. inferior appendages with the first segment very large and sub-quadrangular and the second segment very small and crescent shaped. Phallus and parameres remarkably long and slender.

Genitalia (♀): Not available.

Holotype: INDIA: Assam: NEFA, Kameng Frontier Division; Arunachal Pradesh: Phutang, 3500–7200 ft., 28-IX-1961.

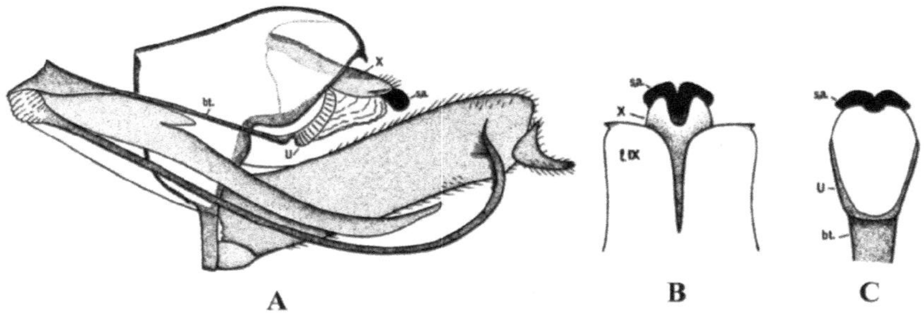

Figure 7.152 *Rhyacophila sherchokpa* (♂): (A) Lateral view of male genitalia; (B) Dorsal view of Xth segment; (C) Tergum, dorsal (from Schmid, 1970).

Holotype depository: ROM, Canada.

Distribution: India (Arunachal Pradesh, Assam).

134. *Rhyacophila sherpa* Schmid

Rhyacophila sherpa, Schmid, 1970: 156–157

Adult: Forewings with golden macules larger and more irregularly arranged than in other species of the group. Wingspan: 14–17.5 mm.

Genitalia (♂): (Figs. 7.153A–B). Vertical part of the Xth segment forming a large horizontal roof, strongly concave downwards and widely indented at its end. The second segment of the inferior appendages does not have its edges aligned with those of the first, very obliquely truncated at its tip. Ventral plate of the phallus long.

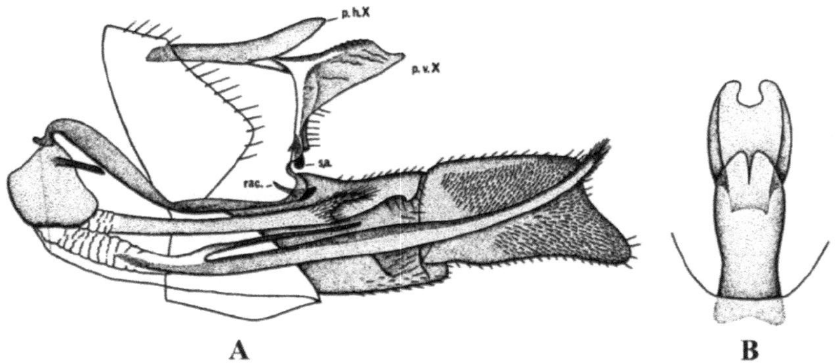

Figure 7.153 *Rhyacophila sherpa* (♂): (A) Lateral view of male genitalia; (B) Dorsal view of Xth segment (from Schmid, 1970).

Genitalia (♀): Not available.

Holotype: INDIA: Sikkim, Tikipchu, 13800 ft., 19-IV-1959.

Holotype depository: ROM, Canada.

Distribution: India (Sikkim).

135. *Rhyacophila shingripa* Schmid

Rhyacophila shingripa, Schmid, 1970: 179

Adult: Top of the body black-brown with concolorous hairiness. Pleura, forelegs and middle reddish-brown. Abdomen and posterior femurs orange-pink; hind tibiae and tarsi dark brown. Uniformly black wings. Wingspan ♂: 15–19 mm.

Genitalia (♂): (Figs. 7.154A–C). Apical dorsal lobe of segment IX forming a large mass. Preanal appendages small and obtuse. Anal sclerites small, slender, vertical and curved backwards. Dorsal appendages of the phallic apparatus very long, quite slender and crossed above the anal sclerites; phallus composed of a dorsal branch split into two slender lobes widely separated from each other and a ventral branch in a long cylinder strongly arched upwards; ventral plate long slender and membranous band; very strong and spiniferous parameres.

Genitalia (♀): Not available.

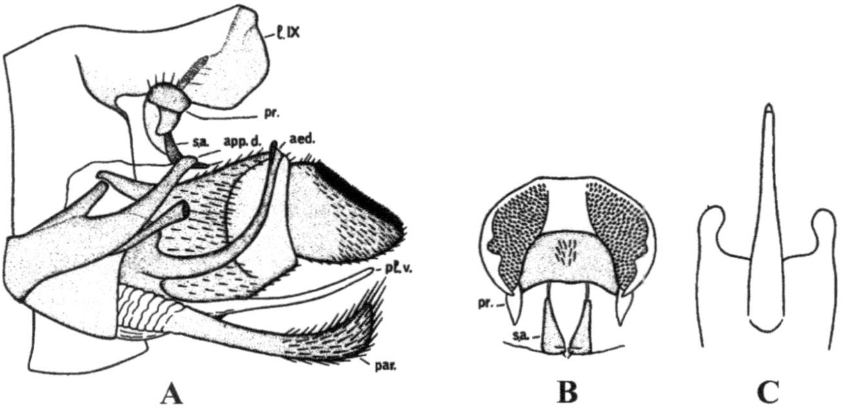

Figure 7.154 *Rhyacophila shingripa* (♂): (A) Lateral view of male genitalia; (B) Front view of Xth segment; (C) Phallic apparatus, dorsal view (from Schmid, 1970).

Holotype: INDIA: Sikkim; Chateng, 6500–8000 ft., 28-VII-1959.

Holotype depository: ROM, Canada.

Distribution: India (Sikkim).

136. *Rhyacophila sikungpa* Schmid

Rhyacophila sikungpa, Schmid, 1970: 167–168

Adult: Top of the body black, with brown hairs. Dark brown pleura and legs. Uniformly black wings. Wingspan: 15–19 mm.

Genitalia (♂): (Figs. 7.155A–B). In dorsal view, the dorsal piece appears obtuse and very slightly indented at its end; ventrally, it shows somewhat swollen apex and a tiny lower tooth. Anal sclerite small, slightly curved and distinctly bilobed. Second segment of the inferior appendages a little longer than high and with its two widely rounded apical angles.

Genitalia (♀): Not available.

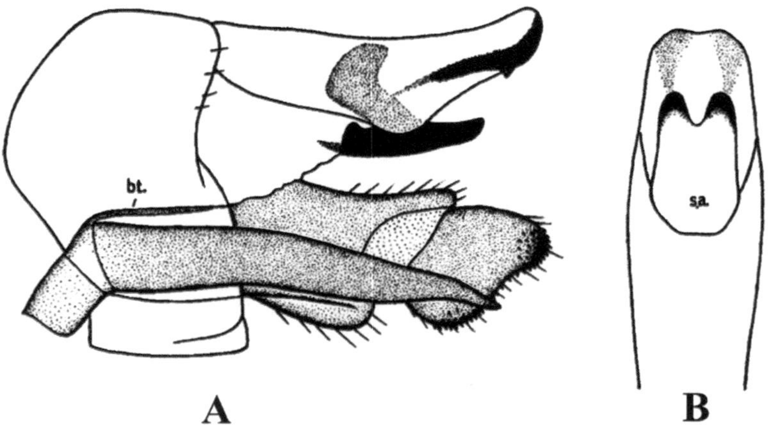

Figure 7.155 *Rhyacophila sikungpa* (♂): (A) Lateral view of male genitalia; (B) Dorsal view of IXth and Xth segments (from Schmid, 1970).

Holotype: INDIA: Uttarakhand; Pauri Garhwal, Akhrotkoti, 5700 ft., 17-18-V-1958.

Holotype depository: ROM, Canada.

Distribution: India (Uttarakhand).

137. *Rhyacophila similis* Martynov

Rhyacophila similis, Martynov, 1935: 99–100

Adult: Head black-brown, with brown antennae; thorax black-brown; legs brownish. Anterior wings brown, shining; apical forks 1 and 2 beginning at the same level, fork 4 much earlier than 3rd. Posterior wings somewhat paler; apical forks 1 and 2 beginning also at the same level. Abdomen brown above, paler, yellowish beneath.

Genitalia (♂): (Figs. 7.156A–C). IXth segment narrow; Xth segment completely united with IXth tergite into a large and thick plate, somewhat dilated in its basal part and excised behind, if seen from above. Behind the plate are situated below two oval blackish bodies, connected by a pale median membrane. Basal joint of inferior appendages extremely broad, if seen laterally, with somewhat convex lower margin; dorsally, the second joint is narrower, but its distal portion is concave, rounded and beset internally with a dense brush of short blackish hairs. Phallus very slender and long.

Genitalia (♀): Not available.

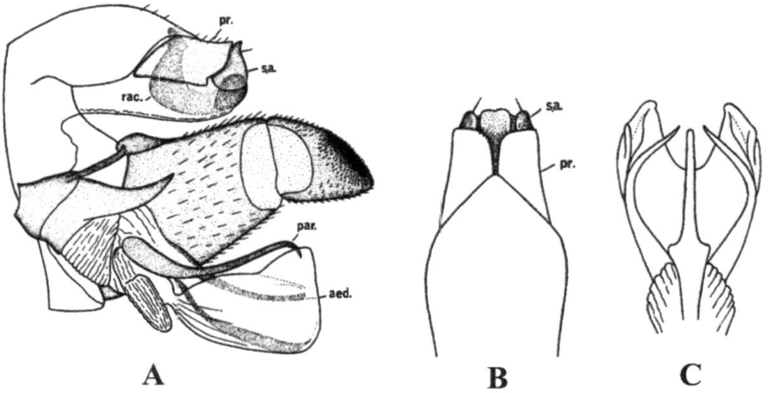

Figure 7.156 *Rhyacophila similis* (♂): (A) Lateral view of male genitalia; (B) Dorsal view; (C) Phallic apparatus, dorsal (from Schmid, 1970).

Holotype: INDIA: Uttarakhand; Jabharket, Dehra, 20-25-VI-1930.

Distribution: Taiwan, India (Uttarakhand).

138. *Rhyacophila spinalis* Martynov

Rhyacophila spinalis, Martynov, 1930: 69–70

Adult: Entire body brown except antenna, labial palp, maxillary palp and legs. Wings pale brown with brown irrotations. Length of antenna 11.64 mm and forewing 14.55 mm. In forewing fork 1st arise earlier than 2nd.

Genitalia (♂): (Figs. 7.157A–C). In dorsal view, dorsal apical lobe of segment IX indented. In lateral view, tergum of segment IX longer than its sternum, anal sclerite truncate shaped apically. Dorsal appendage of phallic apparatus small and hidden beneath segment IX, paramere smaller than phallicata and terminal segment of inferior appendage with serration. In ventral view apices of terminal segment of inferior appendage rounded.

Genitalia (♀): Not available.

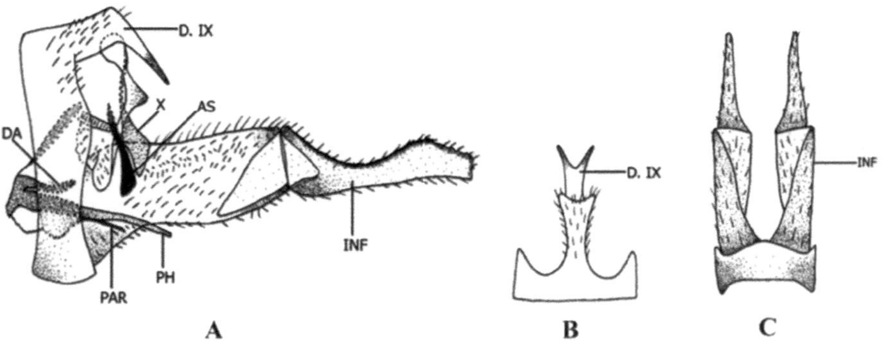

Figure 7.157 *Rhyacophila spinalis* (♂): (A) Lateral view of male genitalia; (B) Dorsal view; (C) Ventral view.

Holotype: TIBET: Gyangtse, 13000 ft., VI-1904.

Holotype depository: BMNH, London.

Additional material examined: INDIA: Meghalaya: Badapani, 800 m, 09-V-2012, 1♂. (Coll. Parey & Pandhar).

Distribution: Nepal, Tibet, Bhutan, India (Arunachal Pradesh, Uttarakhand, Sikkim, Meghalaya).

139. *Rhyacophila stenostyla* Martynov

Rhyacophila stenostyla, Martynov, 1930: 70–71

Adult: Head, thorax, abdomen and wings brown; antenna, legs, maxillary and labial palpi pale brown. Forewing with dark brown irrotation. Length of antenna 9.7 mm and forewing 13.6 mm. In forewing forks 1st and 2nd arise at the same level.

Genitalia (♂): (Figs. 7.158A–C). In dorsal view dorsal apical lobe of segment IX truncate at apex. In lateral view segment IX sub-rectangular in shape, slightly excised at base of inferior appendage. Segment X curved downwards. Anal sclerite slightly hidden beneath segment X, truncate and setiferous apically. Basal segment of inferior appendage long and broader at base; terminal segment about 1/3rd as long to basal segment, but superior lobe rounded and inferior lobe protruding Phallicata broader at base and gradually go on decreasing. Paramere slightly curved near apex and arched upwards apically. In ventral view terminal segment of inferior appendage with concavity at middle.

Genitalia (♀): Not available.

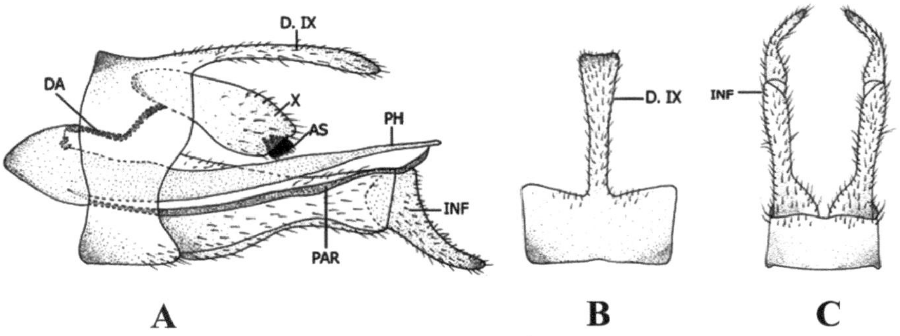

Figure 7.158 *Rhyacophila stenostyla* (♂): (A) Lateral view of male genitalia; (B) Dorsal view; (C) Ventral view.

Holotype: TIBET: Tasam, Rongshar valley, 12000ft., 10-VI-1924.

Holotype depository: BMNH, London.

Additional material examined: INDIA: Uttarakhand: Gobinddham, 3100 m, 20-VI-2011, 1♂. (Coll. Parey & Pandhar).

Distribution: Nepal, Tibet, India (Sikkim, Uttarakhand).

140. *Rhyacophila sumdopa* Schmid

Rhyacophila sumdopa, Schmid, 1970: 168

Adult: Dorsal side of the body black, with brown hairs. Reddish-brown pleura and legs. Forewings uniformly black, with a large golden spot on the arculus. Wingspan: 15–19 mm.

Genitalia (♂): (Figs. 7.159A–B). Dorsal piece with a clearly visible suture, lined with a difference in level between the apical dorsal lobe of the IXth segment and the preanal appendages; the apex has a complex and characteristic shape. Anal sclerite small and curved upwards at its extremity. Sagittal appendage oval and as long as it is wide. Second segment of the inferior appendages as high as long, ending in two regularly rounded lobes, equal and separated by a very small indentation. phallus with a small subapical dorsal tip.

Genitalia (♀): Not available.

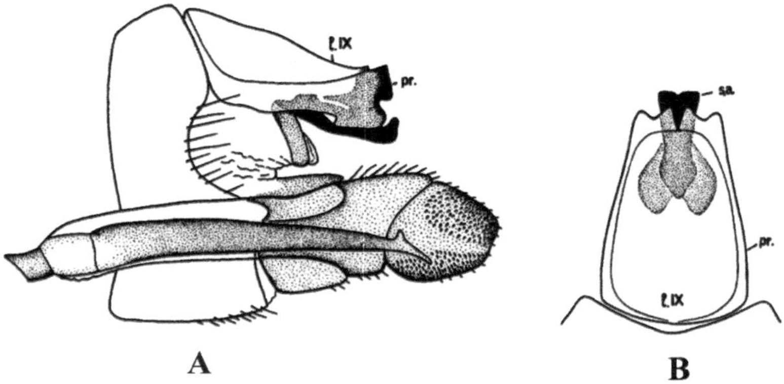

Figure 7.159 *Rhyacophila sumdopa* (♂): (A) Male genitalia, lateral view; (B) Dorsal view (from Schmid, 1970).

Holotype: INDIA: Manipur; Mattiyang, 17-VI-1960.

Holotype depository: ROM, Canada.

Distribution: India (Manipur).

141. *Rhyacophila tarkiya* Schmid

Rhyacophila tarkiya, Schmid, 1970: 194–195

Adult: Head, thorax, abdomen and wings brown; antenna, legs, maxillary and labial palpi pale brown. Wingspan: 19–25 mm.

Genitalia (♂): (Figs. 7.160A–B). Genitalia similar to those of *R. chulukpa* and hardly differing from them except by the phallus, which is shorter, thinner, ends in a much longer slender cylinder and whose ante-apical hairy heels are much less marked (Schmid, 1970).

Genitalia (♀): Not available.

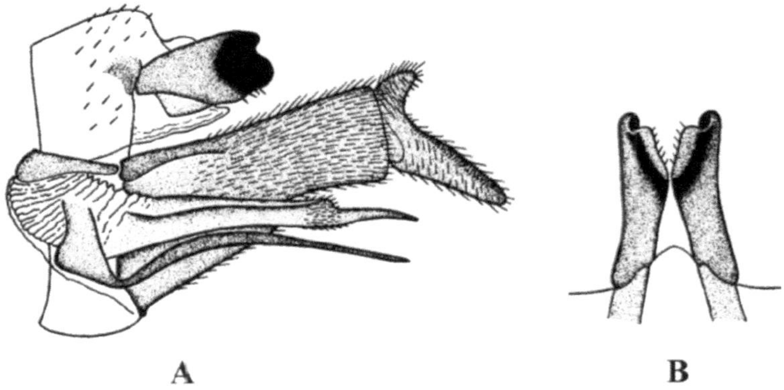

Figure 7.160 *Rhyacophila tarkiya* (♂): (A) Lateral view of male genitalia; (B) Dorsal view of Xth segment (from Schmid, 1970).

Holotype: INDIA: Sikkim; Karponang, 9900–12500 ft., 23-VIII-1959.

Holotype depository: ROM, Canada.

Distribution: India (Sikkim).

142. *Rhyacophila tashepa* Schmid

Rhyacophila tashepa, Schmid, 1970: 143

Adult: Dorsal side of the body dark reddish-brown, with strong golden hairs. Pleura reddish-brown; light red legs. Forewings broad, with a dark brown ground color and riddled with golden macules, anastomosed, and covering a surface superior to that of the ground. Wingspan: 18–21 mm.

Genitalia (♂): (Figs. 7.161A–C). X[th] segment very small and armed with two median points. Second segment of the inferior appendages very large, triangular, very obtuse at the apex and furnished with a narrow peripheral band of spinules. Internal relief of the first segment very pronounced and with a prominent medio-ventral tongue. Very long phallic apparatus.

Genitalia (♀): Not available.

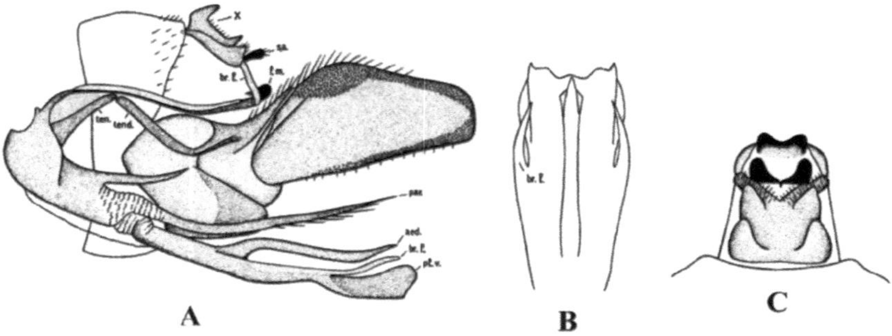

Figure 7.161 *Rhyacophila tashepa* (♂): (A) Lateral view of male genitalia; (B) Dorsal view of phallus; (C) Dorsal view of Xth segment (from Schmid, 1970).

Holotype: INDIA: Sikkim; Chachu, 5700–9500 ft., 21-V-1959.

Holotype depository: ROM, Canada.

Distribution: India (Sikkim).

143. *Rhyacophila tashidingpa* Schmid

Rhyacophila tashidingpa, Schmid, 1970: 195

Adult: Top of the body reddish-brown, with concolorous hairiness. Pleura and legs reddish yellow. Forewings dark brown, intensely mottled with gold. Wingspan: 20.5 mm.

Genitalia (♂): (Figs. 7.162A–C). Preanal appendages appearing triangular when viewed from above and from side. Simple phallus, desclerotized at its base and without parameres. Very long inferior appendages, with the second segment very curiously indented and armed with long spiny bristles at its extremity.

Genitalia (♀): Not available.

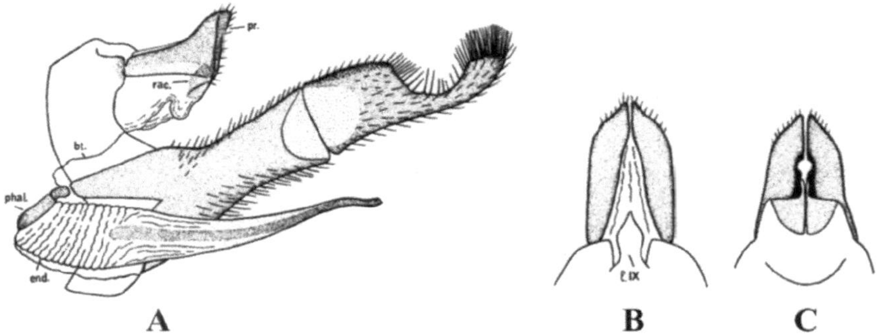

Figure 7.162 *Rhyacophila tashidingpa* (♂): (A) Lateral view of male genitalia; (B) Dorsal view of Xth segment; (C) Dorsal view of Xth segment (from Schmid, 1970).

Holotype: INDIA: Assam; Kameng, NEFA, Kameng Frontier Division, Lungdur, 2800–4900 ft., 11-12-X-1961.

Holotype depository: ROM, Canada.

Distribution: Nepal, India (Arunachal Pradesh).

144. *Rhyacophila tecta* Morton

Rhyacophila tecta, Morton, 1900: 4

Adult: Wings with apex sub-acute; apical forks 1 and 2 sub-equal in the forewings; radius not distinctly furcate at apex. Membrane smoky with brownish pubescence, neuration fuscous. Costal hairs simple. Antennae blackish fuscous, faintly annulated with paler at the joints.

Genitalia (♂): (Figs. 7.163A–C). IXth segment with its lateral margins broadly excised, ventral margin produced in a short, rounded, unpigmented triangle. Xth tergite forming a long, narrow, parallel-sided hood with a rounded apex, from the side very obliquely truncate. Xth sternite membranous. Phallus with a short basal sleeve, simple, slender, slightly bent downwards, its apex with a pair of small, acute spines. No apparent sheaths. Inferior appendage short and stout, extending about as far as apex of Xth tergite. Basal segments with their inner margins fused, terminal segment short, somewhat reniform from the side, inner apical margins set with short, black spines.

Genitalia (♀): Not available.

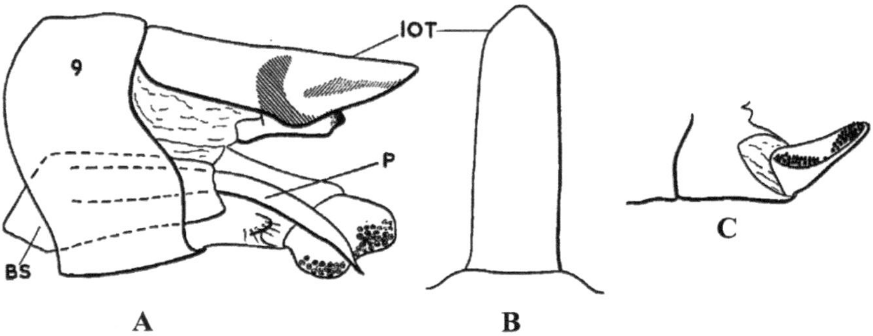

Figure 7.163 *Rhyacophila tecta* (♂): (A) Lateral view of male genitalia; (B) Dorsal view of Xth segment; (C) Ventral view of right inferior appendage (from Kimmins, 1953).

Holotype: INDIA: Assam; Khasi Hills, 6-XII-1899.

Holotype depository: BMNH, London.

Distribution: Taiwan, Bhutan, Myanmar, India (Arunachal Pradesh, Meghalaya).

145. *Rhyacophila tengyelingpa* Schmid

Rhyacophila tengyelingpa, Schmid, 1970: 173

Adult: Dorsal side of the body blackish, with golden hairs. Pleura and legs reddish yellow. Uniformly beige wings. Wingspan: 14–16 mm.

Genitalia (♂): (Figs. 7.164A–B). Preanal appendages and anal sclerites appearing long and low, in lateral view. Phallus slender, simple, arching downwards and armed with a small upper basal point; parameres in broad spines bearing an upper row of setae. Second segment of the inferior appendages angular and with its apical edge very slightly curved.

Genitalia (♀): Not available.

Holotype: INDIA: Sikkim; Zema, 8900–9500 ft., 24-V-1959.

Holotype depository: ROM, Canada.

Distribution: Nepal, India (Arunachal Pradesh).

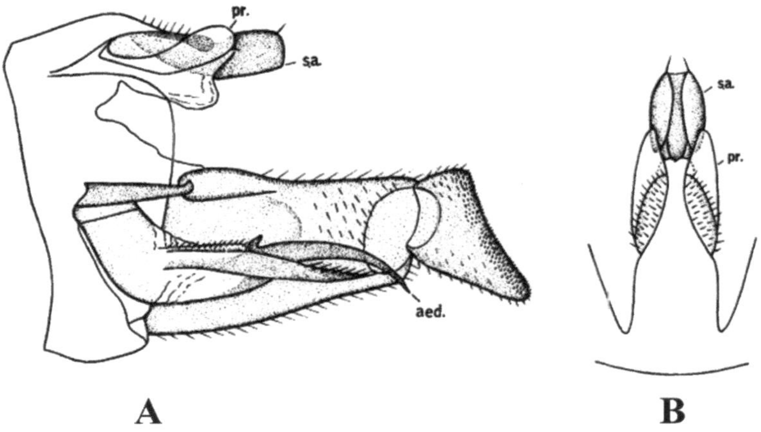

Figure 7.164 *Rhyacophila tengyelingpa* (♂): (A) Male genitalia, lateral view; (B) Dorsal view (from Schmid, 1970).

146. *Rhyacophila tolungpa* Schmid

Rhyacophila tolungpa, Schmid, 1970: 182

Adult: Dorsal side of the body brown black, with black hairs. Pleura and legs brown, but tip of median tarsi, lower half of posterior tarsi and hind femora orange-yellow. Forewings black, with a circular apical patch of gray hairs. Wingspan: 14.5–16 mm.

Genitalia (♂): (Figs. 7.165A–B). In lateral view, the apex of the apical dorsal lobe of segment IX appearing pointed, but not raised. Preanal appendages in elongated ovals. Bifid anal sclerites, with the upper branch adjoining the IXth segment and the lower branch prominent and curved upwards. Lower branch of the phallus as long as the dorsal appendages of the phallic apparatus. Second segment of the inferior appendages small, rounded, but slightly convex.

Genitalia (♀): Not available.

Holotype: INDIA: Assam; NEFA, Kameng Frontier Division, Lifakpo, 1100–7200 ft., 15-III-1961.

Holotype depository: ROM, Canada.

Distribution: Bhutan, India (Arunachal Pradesh).

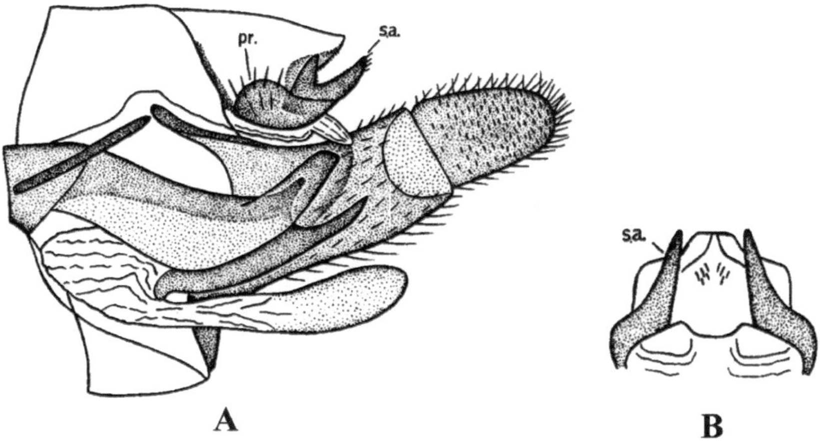

Figure 7.165 *Rhyacophila tolungpa* (♂): (A) Male genitalia, lateral view; (B) Dorsal view of IXth and Xth segments (from Schmid, 1970).

147. *Rhyacophila trashipa* Schmid

Rhyacophila trashipa, Schmid, 1970: 212

Adult: Wings narrow and slender in shape. The forewings have a well-contrasted coloration, with clear and dark areas, well defined, uniform or slightly mottled. Wingspan: 19–24 mm

Genitalia (♂): (Figs. 7.166A–B). Xth segment obtuse at its end and with a conspicuous subapical median tooth. Second segment of the inferior appendages rather elongated, with its apical edge long and straight and its lower edge concave downwards. Lower angle of the first segment approximately 80°. phallus long and accompanied by two membranous lateral lobes, slender, as long as itself and a short ventral lobe, also membranous; robust parameres.

Genitalia (♀): Not available.

Holotype: INDIA: Manipur; Sirohi Kashong, 500–7500 ft., 6-7-VI-1960.

Holotype depository: ROM, Canada.

Distribution: Bhutan, India (Manipur).

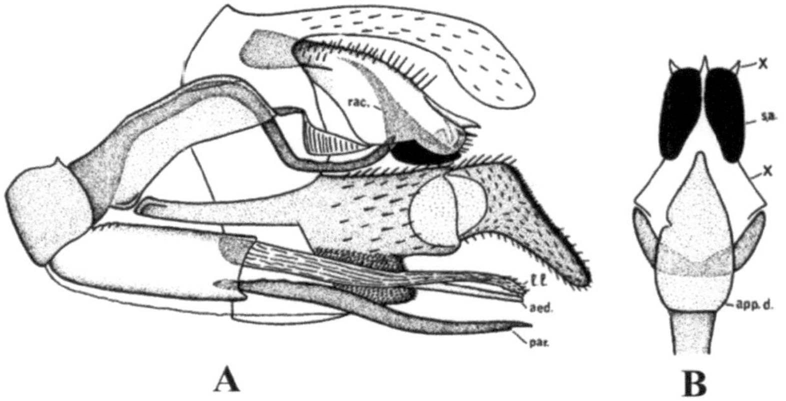

Figure 7.166 *Rhyacophila trashipa* (♂): (A) Male genitalia, lateral view; (B) Tergal assembly and dorsal appendage of phallic apparatus (from Schmid, 1970).

148. *Rhyacophila trulungpa* Schmid

Rhyacophila trulungpa, Schmid, 1970: 202–203

Adult: Dorsal side of the body reddish-brown, with concolorous hairs. Pleura and legs reddish yellow. Narrow wings: the front ones are coarsely and irregularly mottled with brown and light and generally pale in color. Wingspan: 15–23 mm.

Genitalia (♂): (Figs. 7.167A–C). In dorsal view, Xth segment appears narrow, but widened at its end and with a tiny median indentation; in lateral view, none of its apical angles is protruding. Dorsal appendage of the phallic apparatus rather obtuse at its extremity; slender, twisted parameres at apex, somewhat membranous and with a single bunch of spines, in apical position. Second segment of the inferior appendages short and with its two obtuse angles.

Genitalia (♀): Not available.

Holotype: INDIA: Uttarakhand; Pauri Garhwal, Hilang, 5000 ft., 7-VI-1958.

Holotype depository: ROM, Canada.

Distribution: Pakistan, India (Uttarakhand).

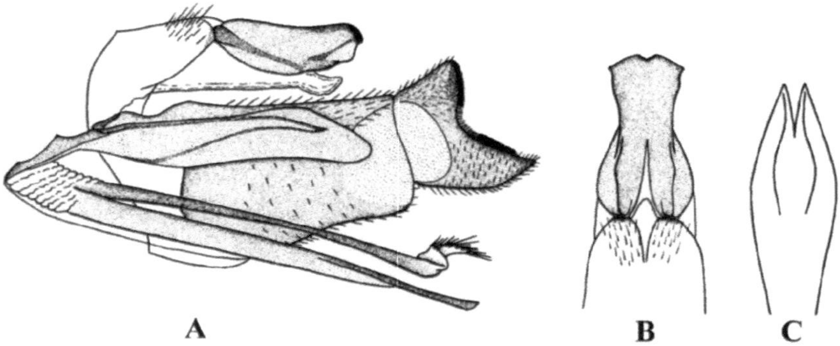

Figure 7.167 *Rhyacophila trulungpa* (♂): (A) Lateral view of male genitalia; (B) Dorsal view; (C) Dorsal appendage of phallic apparatus (from Schmid, 1970).

149. *Rhyacophila tsering* Schmid

Rhyacophila tsering, Schmid, 1970: 148

Adult: Entire body brown except labial palps which are pale brown. Length of antenna 4.85 mm and forewing 5.82 mm. In forewing forks 1st and 2nd arise at the same level.

Genitalia (♂): (Figs. 7.168A–C). In dorsal view, anal sclerite rounded and apices of segment X sclerotized. In lateral view, sternum of segment IX slightly smaller than its tergum. Anal sclerite rounded. Dorsal appendage of phallic apparatus divided into two parts upper part round and broader than lower part and lower part longer than upper part and rod shaped. Paramere rounded apically and upper arm of phallicata rounded and paramere not extending beyond terminal segment of inferior appendage. Basal segment of inferior appendage with deep concavity from superior side near base and terminal segment of inferior appendage with minute spinules on superior as well as on inferior lobes. Superior lobe with superior side convex and inferior lobe pointed apically. In ventral view, basal segment of inferior appendage bulbous.

Genitalia (♀): Not available.

Holotype: INDIA: Kerala; Kandalur, 600–7300 ft., 12-XII-1961.

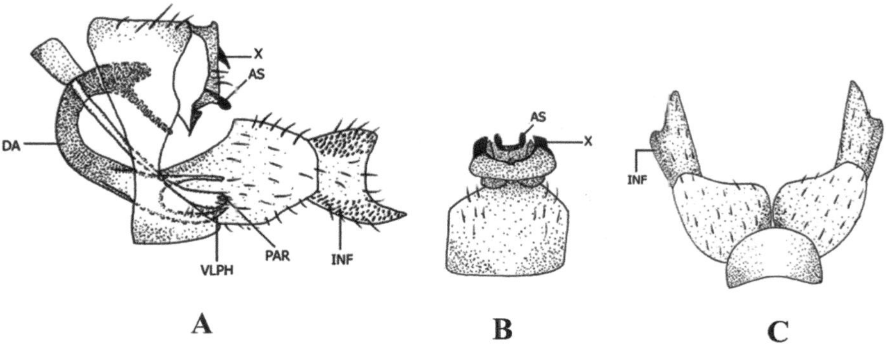

Figure 7.168 *Rhyacophila tsering* (♂): (A) Male genitalia, lateral view; (B) Dorsal view; (C) Ventral view.

Holotype depository: ROM, Canada.

Additional material examined: INDIA: Meghalaya; Bhagmara, 950 m, 12-IV-2009, 1♂. (Coll. Parey & Pandhar).

Distribution: India (Kerala, Meghalaya).

150. *Rhyacophila tshiringpa* Schmid

Rhyacophila tshiringpa, Schmid, 1970: 157

Adult: Upper body dark brown, pleura and legs reddish yellow. Wings black-brown. Wingspan: 18–26 mm.

Genitalia (♂): (Figs. 7.169A–C). Vertical part of the Xth segment inserted at the end of the horizontal part and provided with an odd, long superior point, whose dorsal face is slightly concave and is broader at its end than at its base, lower tip absent. Inferior appendages not very long; second segment longer than high, provided with a deep apical indentation which isolates two lobes: an upper, triangular and a lower, longer, and slender.

Genitalia (♀): Not available.

Holotype: INDIA: Sikkim; Talam, 5000–12150 ft, 16-VI-1959.

Holotype depository: ROM, Canada.

Distribution: Nepal, India (Sikkim).

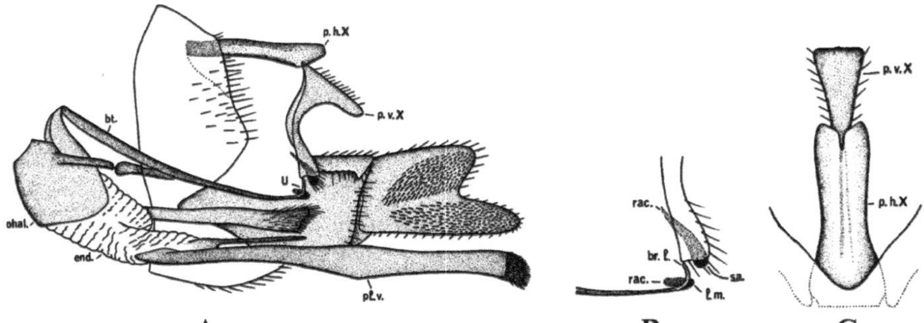

Figure 7.169 *Rhyacophila tshiringpa* (♂): (A) Male genitalia, lateral view; (B) Lower part of Xth segment and tergal assembly, lateral view; (C) Dorsal view of Xth segment (from Schmid, 1970).

151. *Rhyacophila tshogpa* Schmid

Rhyacophila tshogpa, Schmid, 1970: 213

Adult: Wings moderately broad; the forewings bear the usual light and dark zones of the group, but the pattern is mostly uniform and slightly mottled and fused colors. Wingspan: 17–25 mm.

Genitalia (♂): (Figs. 7.170A–C). Apical dorsal lobe of segment IX not exceeding superior apical angle of first segment of inferior appendages. End of segment X without subapical midline tooth. First segment of the inferior appendages shorter, deeply, and widely incised, although the lower tip is long and more robust. Phallus slender and gradually thinning to the end; spiniform parameres, very slender, sometimes bifurcated before their extremity, as long as the phallus and inserted very close to the base of the later.

Genitalia (♀): Not available.

Holotype: INDIA: Sikkim; Rapham, 2-IV-1959.

Holotype depository: ROM, Canada.

Distribution: India (Sikkim).

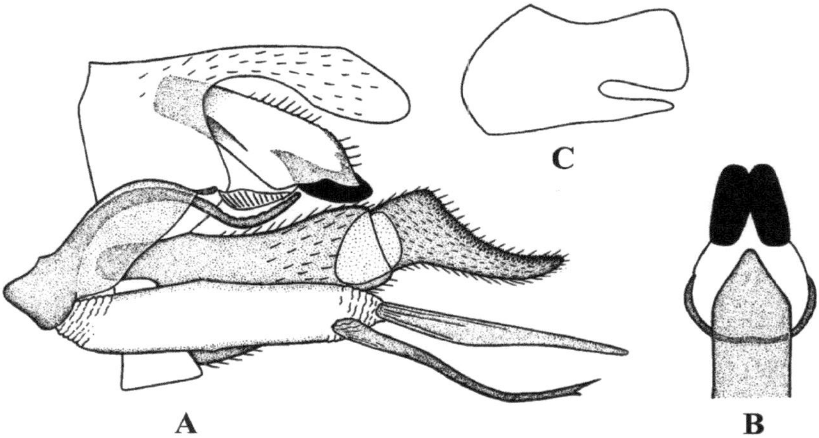

Figure 7.170 *Rhyacophila tshogpa* (♂): (A) Lateral view of male genitalia; (B) Dorsal appendage of phallic apparatus; (C) 1st segment of inferior appendage, lateral (from Schmid, 1970).

152. *Rhyacophila tsiudmarpo* Schmid

Rhyacophila tsiudmarpo, Schmid, 1970: 169

Adult: Dorsal side of the body dark brown, with concolorous hairiness. Dark reddish-brown pleura and legs. Wings uniformly black-brown. Wingspan: 17–19 mm.

Genitalia (♂): (Figs. 7.171A–B). Remarkable for the elongation of the appendages and the dynamism of their forms. Dorsal piece very long, gradually tapering to tip, which is broadly bifid. Anal sclerite in plate much larger than the dorsal part, concave upwards and truncated at its end. Second segment of the inferior appendages in simple crescent and with acute points. Simple and very long phallus.

Genitalia (♀): Not available.

Holotype: INDIA: Assam; NEFA, Kameng Frontier Division, Jhum La, 1-2-VI-1961.

Holotype depository: ROM, Canada.

Distribution: India (Arunachal Pradesh).

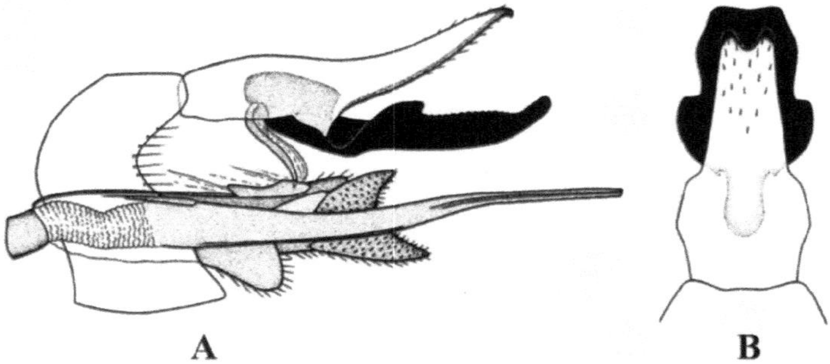

Figure 7.171 *Rhyacophila tsiudmarpo* (♂): (A) Lateral view of male genitalia; (B) Dorsal view (from Schmid, 1970).

153. *Rhyacophila tsona* Schmid

Rhyacophila tsona, Schmid, 1970: 192

Adult: Top of the body yellow-brown, with concolorous hairiness. Fairly light reddish-brown pleura and legs. Forewings brown, mottled with areas of small golden macules. Wingspan: 20–25 mm.

Genitalia (♂): (Figs. 7.172A–B). In lateral view, the preanal appendages and anal sclerites appear almost entirely fused, being distinct and free only at their tip; in dorsal view, the anal sclerites appear very obliquely truncated and terminated in an acute point. Parameres a little thickened in a club at the apex and quite long and hairy. Second segment of the inferior appendages with the upper lobe broadly and regularly rounded and the lower lobe long and parallel-sided.

Genitalia (♀): Not available.

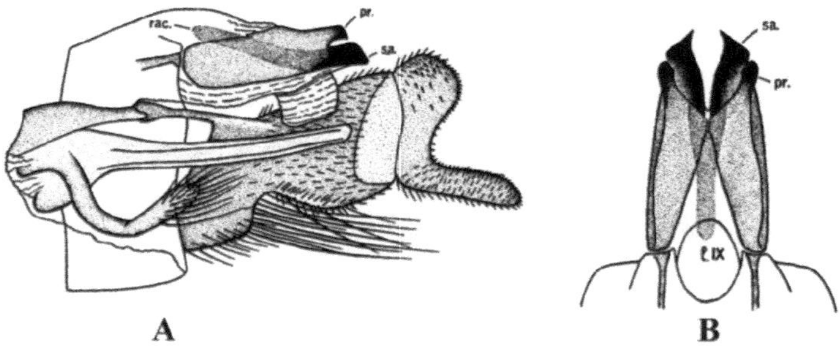

Figure 7.172 *Rhyacophila tsona* (♂): (A) Lateral view of male genitalia; (B) Dorsal view (from Schmid, 1970).

Holotype: INDIA: Manipur, Sirohi Kashong, 5300 ft., 6-7-VI-1960.

Holotype depository: ROM, Canada.

Distribution: Bhutan, India (Manipur).

154. *Rhyacophila tsongkhapa* Schmid

Rhyacophila tsongkhapa, Schmid, 1970: 219

Adult: Body reddish-brown, slightly lighter below. Anterior claws not very large, but strongly asymmetrical; the external claw is small and strongly curved. Narrow and elongated wings. The costal rib of the forewings bears a fringe of bristly hairs, long especially in the middle of the length of the wing. The forewings are yellow, riddled with clearly visible brown macules, forming a rather complex network. Wingspan: 23–32 mm.

Genitalia (♂): (Figs. 7.173A–B). Apical dorsal lobe of the IXth segment clearly flared at its extremity and with its apical lateral points prominent. Xth segment appearing as long as high in lateral view, with its rounded corners. Second segment of the inferior appendages with the upper lobe quite acute and the lower lobe very elongated and ogival in shape. The edges of the indentation bear a single row of fine spines.

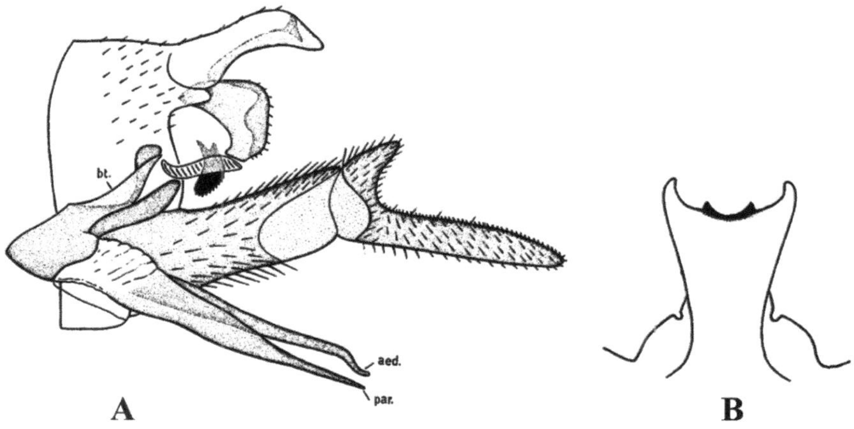

Figure 7.173 *Rhyacophila tsongkhapa* (♂): (A) Lateral view of male genitalia; (B) Dorsal apical lobe of IXth segment, dorsal view (from Schmid, 1970).

Genitalia (♀): (Figs. 7.174A–B). All abdominal segments bear short setae each inserted on a dark circular spot. Valves of the VIIIth segment slightly sclerotic, of simple form and not reaching the dorsal surface of the abdomen. Copulatory bursa terminated in a slightly chitinous stylet.

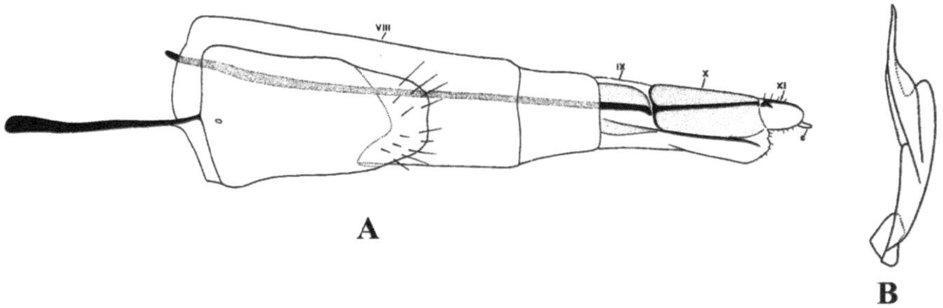

Figure 7.174 *Rhyacophila tsongkhapa* (♀): (A) Lateral view of female genitalia; (B) Copulatory bursa, lateral view (from Schmid, 1970).

Holotype: INDIA: Sikkim, Tangshing, 12000–13000 ft., 15-IV-1959.

Holotype depository: ROM, Canada.

Distribution: India (Sikkim).

155. *Rhyacophila tungkorpa* Schmid

Rhyacophila tungkorpa, Schmid, 1970: 175

Adult: Wingspan: 13–14 mm.

Genitalia (♂): (Figs. 7.175A–C). Trapezoidal preanal appendages. Anal sclerites higher than long; their root extends forward to the level of the beginning of the preanal appendages. Superior appendages of the phallic apparatus quite pointed. Phallus shorter than its ventral plate; very thin parameres. Second segment of the inferior appendages not very high and fairly regularly attenuated.

Genitalia (♀): Not available.

Holotype: INDIA: Sikkim, Tikjak, 2300–5200 ft., 7-IV-1959.

Holotype depository: ROM, Canada.

Distribution: India (Sikkim).

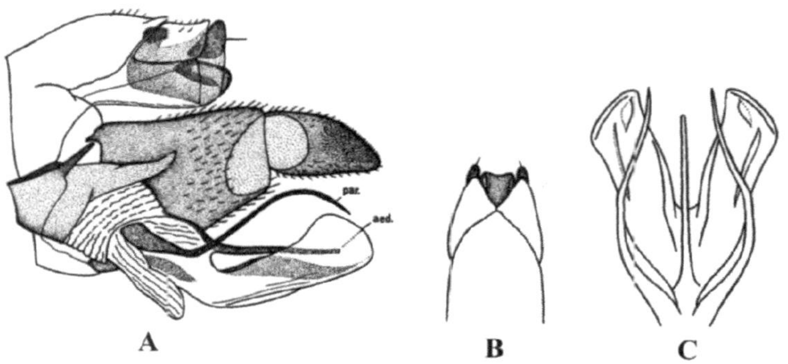

Figure 7.175 *Rhyacophila tungkorpa* (♂): (A) Lateral view of male genitalia; (B) Dorsal view; (C) Phallic apparatus, dorsal view (from Schmid, 1970).

156. *Rhyacophila tungpa* Schmid

Rhyacophila tungpa, Schmid, 1970: 179–180

Adult: Top of the body reddish-brown, with concolorous hairiness. Pleura, legs and abdomen reddish gray brown. Forewings dark brown, riddled with golden spots, quite large and clearly visible. Wingspan: 16–17 mm.

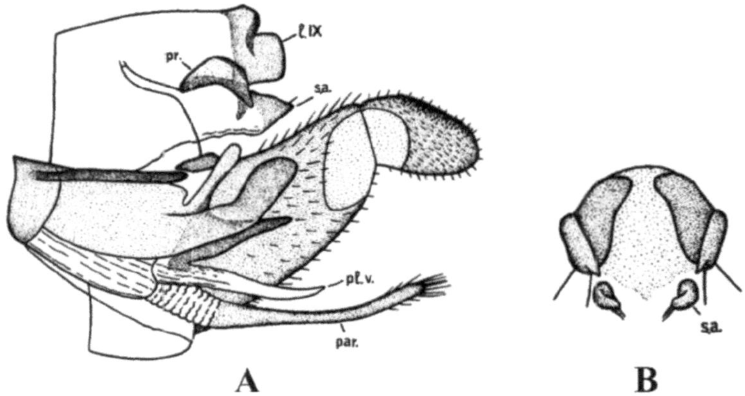

Figure 7.176 *Rhyacophila tungpa* (♂): (A) Lateral view of male genitalia; (B) Front view of Xth segment (from Schmid, 1970).

Genitalia (♂): (Figs. 7.176A–B). Apical dorsal lobe of segment IX forming two obtuse lateral lobes. Preanal appendages in the form of circumflex accents. Small, conical, and prominent anal sclerites. Dorsal appendages of the phallic apparatus strong and rounded; aedeagus composed of a weakly sclerotized and oval dorsal branch and a strongly chitinous and subcylindrical ventral branch; ventral plate slightly sclerotized and long oval. Second segment of the inferior appendages small and rounded.

Genitalia (♀): Not available.

Holotype: INDIA: Uttarakhand; Pauri Garhwal, Jungal Chatti, 11-V-1958.

Holotype depository: ROM, Canada.

Distribution: India (Uttarakhand).

157. *Rhyacophila ugyenpa* Schmid

Rhyacophila ugyenpa, Schmid, 1970: 204–205

Adult: Dorsal side of the body dark brown, with concolorous hairiness. Reddish-brown pleura and legs; gray brown tarsi. Forewings uniformly dark brown. Wingspan: 11.5–14 mm.

Genitalia (♂): (Figs. 7.177A–B). Apical dorsal part of segment IX with scanty brush of hairs. Dorsal part of the Xth segment broadly bifid at its extremity. Parameres long, very slender, and distinctly broadened at their extremity. Second segment of the inferior appendages with tiny apical dorsal lobe and very elongated ventral lobe.

Genitalia (♀): Not available.

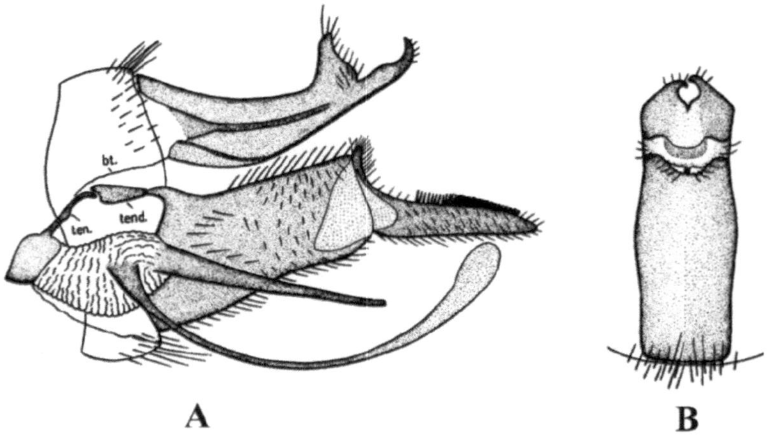

Figure 7.177 *Rhyacophila ugyenpa* (♂): (A) Male genitalia, lateral view; (B) Dorsal view (from Schmid, 1970).

Holotype: INDIA: Assam, United Jaintia and Khasi Hills, Serrarim, 4-X-1960.

Holotype depository: ROM, Canada.

Distribution: India (Meghalaya).

158. *Rhyacophila wangpo* Schmid

Rhyacophila wangpo, Schmid, 1970: 143

Genitalia (♂): (Figs. 7.178A–C). X^th^ segment massive and triangular, with its protruding superior apical angles. Very high inferior appendages, with the second segment of considerable size, notched at its extremity, ending in two lobes of unequal size and armed with internal spinules arranged in an incompletely peripheral band. Phallic apparatus short.

Genitalia (♀): Not available.

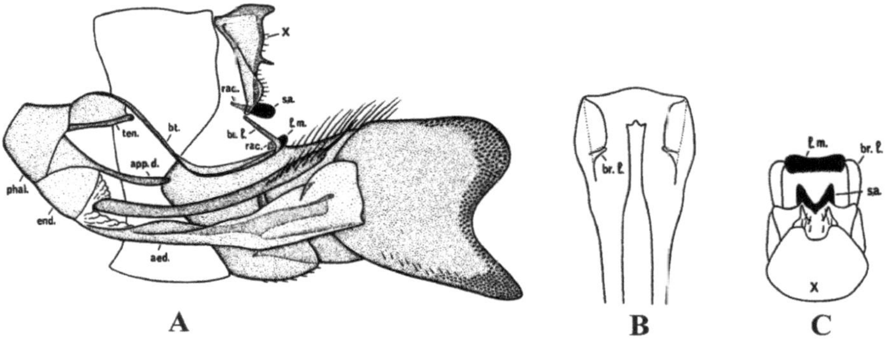

Figure 7.178 *Rhyacophila wangpo* (♂): (A) Lateral view of male genitalia; (B) Dorsal view of phallus; (C) Dorsal view of Xth segment (from Schmid, 1970).

Holotype: INDIA: West Bengal; Khani, 2050 ft., 10-IX-1959.

Holotype depository: ROM, Canada.

Distribution: India (West Bengal).

159. *Rhyacophila yarlungpa* Schmid

Rhyacophila yarlungpa, Schmid, 1970: 182–183

Adult: Upper body dark reddish-brown, with black hair. Light reddish-brown pleura and legs. Uniformly black wings. Rufous abdomen; genital segments orange-yellow. Wingspan: 12–17 mm.

Genitalia (♂): (Fig. 7.179). Apical dorsal lobe of segment IX strongly reduced. Preanal appendages small, sub-quadrangular and with rounded angles. Anal sclerites reduced to a minute point, but their root is developed into a large hollow cylinder, extending into segment VIII. Phallus long, slender, with its upper branch split into two long slender appendages. Inferior appendages equally long and slender in shape, with the second segment ending in two unequal lobes.

Genitalia (♀): Not available.

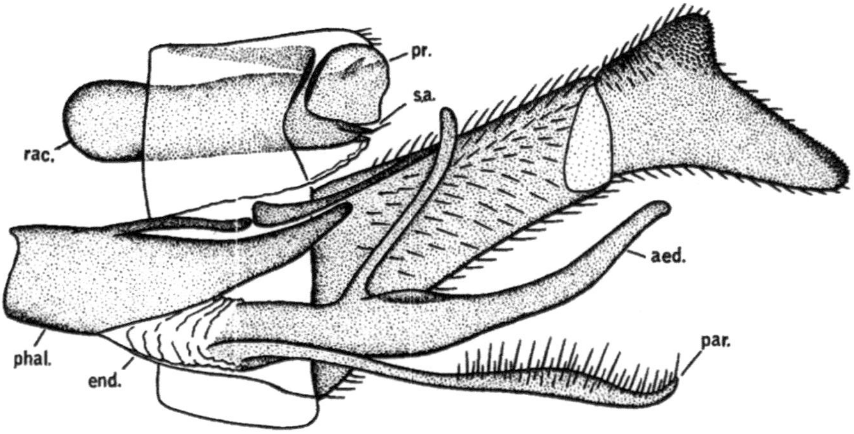

Figure 7.179 *Rhyacophila yarlungpa* (♂): Lateral view of male genitalia (from Schmid, 1970).

Holotype: INDIA: Assam; NEFA, Kameng Frontier Division, Lifakpo, 15-III-1961.

Holotype depository: ROM, Canada.

Distribution: Nepal, India (Arunachal Pradesh).

160. *Rhyacophila yigrongpa* Schmid

Rhyacophila yigrongpa, Schmid, 1970: 141–142

Adult: Top of the body blackish, with concolorous hairiness. Dark brown pleura and legs. Fairly bright orange-yellow abdomen, but with the apical half of the inferior appendages brownish. Forewings uniformly blackish. Wingspan: 15–20 mm.

Genitalia (♂): (Figs. 7.180A–D). Apical dorsal lobe of the IXth segment with a wide oval indentation, isolating two acute lateral points. Inferior appendages long and slender, with the second segment in a slender triangle. Dorsal appendage of the phallic apparatus narrow and truncated at its extremity; parameres with a few sparse apical setae.

Genitalia (♀): Not available.

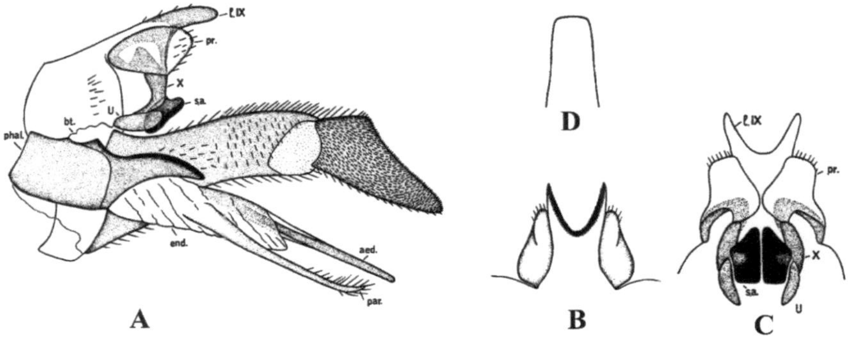

Figure 7.180 *Rhyacophila yigrongpa* (♂): (A) Lateral view of male genitalia; (B) Dorsal view; (C) Front view of Xth segment; (D) Apex of dorsal appendage of phallic apparatus, dorsal view (from Schmid, 1970).

Holotype: INDIA: Sikkim; Bongteng, 6-IV-1959.

Holotype depository: ROM, Canada.

Distribution: India (Sikkim).

161. *Rhyacophila yipung* Schmid

Rhyacophila yipung, Schmid, 1970: 142

Adult: Upper body dark brown. Brownish yellow pleura and legs. Forewings dark brown, with golden, small, scanty and indistinct macules. Wingspan: 15–19 mm.

Genitalia (♂): (Figs. 7.181A–D). Dorsal appendage of the phallic apparatus regularly rounded at its extremity. Upper edge of first segment of inferior appendages very regularly arched; second segment small and triangular. Parameres early divided into two parallel branches, one spiniform, the other spiniferous.

Genitalia (♀): Not available.

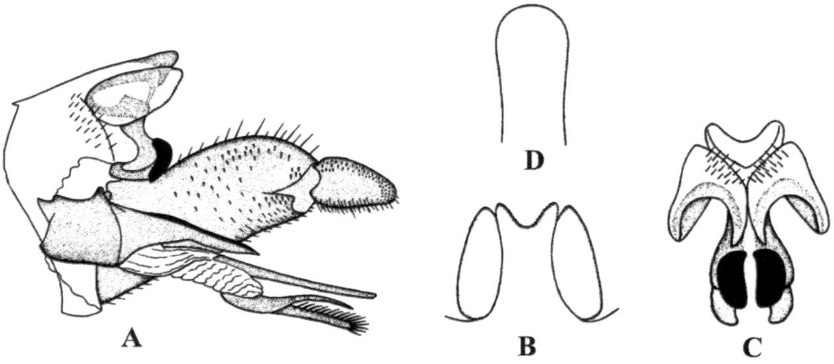

Figure 7.181 *Rhyacophila yipung* (♂): (A) Lateral view of male genitalia; (B) Dorsal view; (C) Front view of Xth segment; (D) Apex of dorsal appendage of phallic apparatus, dorsal view (from Schmid, 1970).

Holotype: INDIA: Manipur; Mattiyang, 2500–5000 ft., 17-VI-1960.

Holotype depository: ROM, Canada.

Distribution: India (Manipur).

162. *Rhyacophila yishepa* Schmid

Rhyacophila yishepa, Schmid, 1970: 177–178

Adult: Dorsal side of the body black, with concolorous hairiness. Dark brown pleura and legs. Black wings, with a very small discolored area covered with gray hairs, at the apex of the forewings. Bright orange abdomen. Wingspan: 14–15 mm.

Genitalia (♂): (Figs. 7.182A–C). Preanal appendages very similar in shape to those of *Rhyacophila churongpa*, but with more prominent apical point and narrower indentation. Dorsal appendages of the phallic apparatus distinctly slender; lateral lobes of the phallus small, narrow, not very prominent, and unarmed; ventral plate forming two spikes distinctly longer than phallus. Inferior appendages regularly narrowed from base to tip and with second joint sub-quadrangular.

Genitalia (♀): Not available.

Holotype: INDIA: Sikkim; Yoksam, 4600–8100 ft., 30-IX-1959.

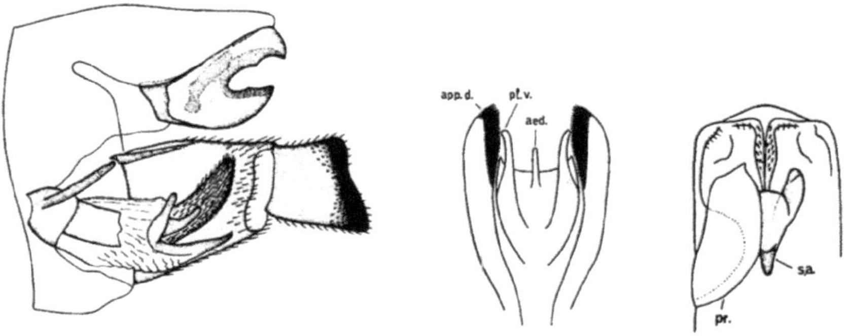

Figure 7.182 *Rhyacophila yishepa* (♂): (A) Lateral view of male genitalia, with dorsal appendage of left phallic apparatus omitted; (B) Ventral view of phallic apparatus; (C) Front view of Xth segment, with right preanal appendage omitted (from Schmid, 1970).

Holotype depository: ROM, Canada.

Distribution: India (Sikkim).

163. *Rhyacophila yonggyapa* Schmid

Rhyacophila yonggyapa, Schmid, 1970: 203

Adult: Dorsal side of the body reddish-brown, with fine light golden hairs. Pleura and legs reddish yellow. Wide wings: the front ones are densely pubescent, brown and riddled with clearly visible golden macules. Wingspan ♂: 17 mm.

Genitalia (♂): (Figs. 7.183A–C). Dorsal part of the Xth segment raised and indented dorsally in a triangle at its extremity. In dorsal view, Upper appendage of the phallic apparatus appears long indented and trilobed, reduced parameres. Second segment of the inferior appendages with a notch at right angles isolating two obtuse lobes.

Genitalia (♀): Not available.

Holotype: INDIA: Manipur; Singkhap, 3800 ft., 16-17-VIII-1960.

Holotype depository: ROM, Canada.

Distribution: India (Manipur).

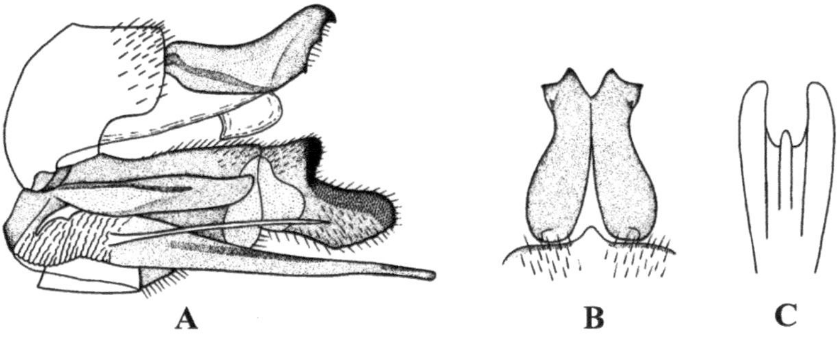

Figure 7.183 *Rhyacophila yonggyapa* (♂): (A) Lateral view of male genitalia; (B) Dorsal view; (C) Dorsal appendage of phallic apparatus, dorsal view (from Schmid, 1970).

164. *Rhyacophila yullha* Schmid

Rhyacophila yullha, Schmid, 1970: 195

Adult: Upper body light brown with colored hairs. Antennae brown with light bands. Wings uniformly blackish.

Genitalia (♂): (Figs. 7.184A–B). Preanal appendages very long, strong and rounded at their extremity. Phallus slightly sclerotized and bifid; parameres in membranous cylinders, erectile but not pleated, twice as long as the phallus and piliferous at their end. First article of the inferior appendages very high at its base, then abruptly thinned by half; second segment small, long, thin and acute.

Genitalia (♀): Not available.

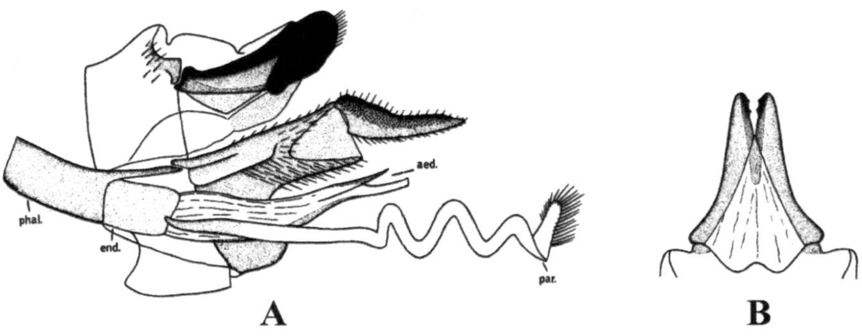

Figure 7.184 *Rhyacophila yullha* (♂): (A) Lateral view of male genitalia; (B) Dorsal view of Xth segment (from Schmid, 1970).

Holotype: INDIA: Sikkim; Gey, 11500–12000 ft., 18-V-1959.

Holotype depository: ROM, Canada.

Distribution: Bhutan, India (Sikkim).

165. *Rhyacophila zhungpa* Schmid

Rhyacophila zhungpa, Schmid, 1970: 199

Adult: Head, antenna, maxillary palps and labial palps pale brown; thorax, abdomen brown; legs and wings orangish. Length of antenna 9.7 mm and forewing 7.76 mm. In forewings, fork 2nd arises slightly earlier than 1st.

Genitalia (♂): (Figs. 7.185A–C). In dorsal view, apices of preanal appendage not fused. Segment IX broad in lateral aspect; dorsal apical lobe of segment IX little prominent. Inferior appendage short, basal segment longer than terminal, terminal segment apically widely excised with superior lobe short and rounded whereas inferior lobe long and triangular. Preanal appendage long and sub- rectangular. Phallicata long with two spines at middle of superior side and one apical spine. Parameres pointed apically with inferior branch which is with serrated margin. In ventral view basal segment of inferior appendage with indentation apically.

Genitalia (♀): Not available.

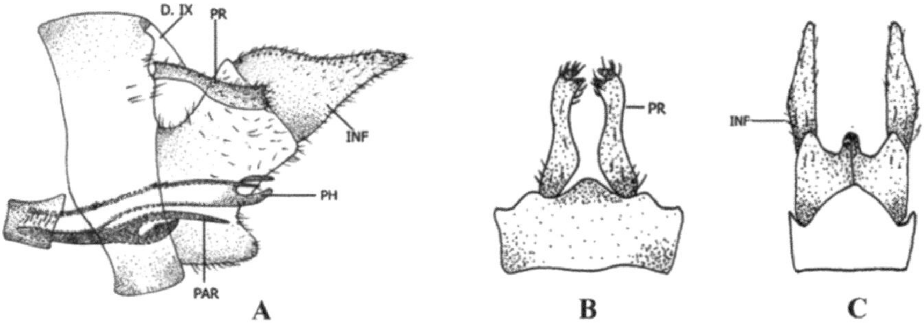

Figure 7.185 *Rhyacophila zhungpa* (♂): (A) Lateral view of male genitalia; (B) Dorsal view; (C) Ventral view.

Holotype: INDIA: Assam; NEFA, Kameng Frontier Division, Amatulla, 11-III-1961.

Holotype depository: ROM, Canada.

Additional material examined: INDIA: Himachal Pradesh: Panjpulla, 2100 m, 12-VIII-2011; Arunachal Pradesh: Dirang, 1700 m, 06-X-2010; Hunli, 1450 m, 02-V-2011; Sikkim: Mangan, 1900 m, 13-V-2009; Sangkalang, 1700 m, 14-V-2009. (Coll. Parey & Pandhar).

Distribution: Bhutan, Nepal, India (Himachal Pradesh, Arunachal Pradesh, Sikkim).

Chapter 8

Ecological Services of Trichoptera (Rhyacophilidae)

Insects are key representatives of ecological biodiversity. A large percentage of insect species inhabit freshwater habitats such as swamps, ponds, lakes, springs, streams, and rivers. They serve a crucial ecological role in the proper functioning of freshwater habitats. Due to their high abundance, high birth rate with a short generation period, vast biomass, and rapid colonization of freshwater habitats, aquatic insects are recognized as model species in assessing the structure and function of the freshwater ecosystem. Mayflies (Ephemeroptera), dragonflies (Odonata), and caddisflies are some of the well-known aquatic insects of inland wetlands (Trichoptera). Caddisflies are the most important category of aquatic insects that contribute significantly to ecosystem services. Trichoptera provides ecological services like critical roles in food webs, biological water quality monitoring, and food for fishes and other predators of human relevance (Morse, 2017). Their silk is providing a paradigm for interesting prospective applications in medicine, material science, and textiles, and they have been credited with engineering their habitats and resolving forensic questions. Some of the essential services of Trichoptera are discussed hereunder:

Rhyacophilidae of India
Sajad Hussain Parey, Tabraq Ali, and Manpreet Singh Pandher

ISBN 978-981-4968-74-4 (Hardcover), 978-1-003-41095-9 (eBook)
www.jennystanford.com

8.1 Biomonitoring Programs

For many years, living creatures in freshwater habitats have served as sentinels against water contamination, particularly organic pollutants. Toxicological assessments of selected model species' responses to varying amounts of various effluents, for example, have been the focus of controlled trials (Klaassen, 2019). Changes in the populations or communities in rivers have been studied by other projects. Most commonly, changes in fish populations or communities, algae, and especially macroinvertebrates, have been exploited for this purpose (Barbour et al., 1999). Caddisflies, stoneflies (Plecoptera), and Mayflies (Ephemeroptera) have been proven to be less tolerant to organic and other types of pollution than other freshwater taxa, hence, changes in their abundance and diversity are frequently highlighted in freshwater biomonitoring programs.

8.2 Contribution to the Betterment of Aquatic Ecosystems

Caddisflies are an essential component of the diet of freshwater fishes. Trichoptera larvae are a food source for bottom-feeding fish in a variety of freshwater bodies. Trichoptera larvae and adults are occasionally eaten by predatory juvenile fishes. The macro caddisflies, which come in large numbers such as Limnephilidae and Phryganeidae species, are the most useful meal for the fish. Caddisflies are a favorite diet of terrestrial and wild aquatic birds, as well as several mammals, especially when they flock together. When these insects are resting on trees and stones, amphibians like frogs and reptiles like snakes and lizards, as well as mammals like bears, feed on them (Hickin, 1967).

Caddisfly larvae perform a crucial part in keeping their aquatic environment clean. Caddisflies that make cases tear up plants for the creation of their cases, eliminating organic detritus from the habitats and niches they occupy indirectly. They eat practically any type of organic stuff found in water bodies, whether alive or dead. Herbivorous species eat only plants, whereas carnivorous species eat a variety of freshwater animals and engage in cannibalism within their own species (Wissinger et al., 2004).

8.3 Trichoptera as Aquatic Pollution Indicators

Caddisflies have the credibility to be used as a water pollution indicator because of the huge and diverse group they represent (Wiggins, 1996). Caddisfly larvae are mostly found in cool and lotic environments, but they can also be found in warm and lentic environments. They are site-specific, and some groups can adapt to a more hostile environment. The accessible niche in the habitat can be used more efficiently because of these adjustments.

Caddisfly larvae are noted for building retreats, nets, and portable cases or tubes. These case-making activities are intertwined with their various roles in the environment. The cases not only provide a safe haven for the larvae but also aid them in capturing or obtaining food particles from the current via the nets. Portable cases allow the larvae to move from one location to another in search of food, and they also improve the effectiveness of the larvae's respiration by allowing them to create artificial currents rather than relying on natural currents (Wiggins, 1996). Caddisflies are frequently found to be more plentiful in species and biologically more heterogeneous in any given habitat. As a result of their adaptability to various habitats and case-making tendencies, caddisfly larvae have the potential to become one of the most dependable markers of water contamination (Dohet, 2002).

The inspection of water plays an important role in determining the number of benthic macroinvertebrate species in a population. The pH and DO level of a waterbody are crucial elements for determining the number of caddisflies present at a given location. Caddisfly density is negatively correlated with the alkalinity and hardness of the water. As a result, the lowest concentration of these characteristics may lead to an increase in species abundance (Dani Benchamin et al., 2020).

Table 8.1 Relationship between hydrologic parameters and the density of Trichoptera (Dani Benchamin et al., 2020)

Limnological parameters	Hydropsychidae sp.	Philopotamidae sp.	Rhyacophilidae sp.	Psychomyiidae sp.	Limnephilidae sp.
Water temperature (°C)	−0.333	−0.3088	−0.4104	−0.4289	−0.4154
pH	−0.4866	−0.4862	−0.5033	−0.468	−0.5473
DO (mg/l)	0.4193	0.4385	0.3238	0.3364	0.2638
CO_2 (mg/l)	−0.8332	−0.8415	−0.798	−0.7716	−0.8054
Conductivity (NTU)	−0.1729	−0.1401	−0.3049	−0.3095	−0.3541
Alkalinity (mg/l)	−0.8437	−0.8349	−0.8629	−0.8736	−0.8486
Hardness (mg/l)	−0.6762	−0.6814	−0.6606	−0.6409	−0.6709
K (mg/l)	−0.8161	−0.8214	−0.7944	−0.7659	−0.8124
Na (mg/l)	−0.8117	−0.8167	−0.7904	−0.7619	−0.8086

The density of caddisflies has a considerable positive relationship with DO (Table 8.1). Because of the increased levels of DO in some areas, the diversification of caddisflies is at its peak. Caddisfly density is often low in midstream and downstream sites, where the DO is lower. The abundance of caddisflies is greatest upstream, owing to favorable environmental conditions, the availability of planks, stones, and various substrates at the bottom, and shallow depth (Leprieur et al., 2008). The caddisflies are mostly nonexistent in locations with more depth, allowing filthy water to mix downstream. Meanwhile, increased anthropogenic interference, such as fishing vessels in rivers, waste disposal, civil infrastructure, and different industrial establishments, results in a lack of a suitable substratum for the Trichoptera population (Arnab et al., 2018).

Chapter 9

Threats to and Conservation of Trichoptera

Threats to aquatic insect diversity vary from country to country, and also between the temperate zone and the tropics. Trichoptera, like other aquatic insects, are also vulnerable to a variety of anthropogenic factors. Because of their habitat under the water, these insects are extremely sensitive to water quality. In fact, the diversity of species present in a stream can serve as an indicator of its health, and scientists can use a biodiversity index to track stream quality. Caddisflies, for example, are known to have a low tolerance for poor water quality, therefore their presence signals that the stream is healthy. However, if the stream only includes worm-like organisms and fly larvae, it may be polluted, as these animals have a strong tolerance for poor water quality. Dumping of trash near the streams, runoff waters from areas with improper drainage, city storm drains where people dump a variety of liquids and changing habitats around the stream are all issues that can affect water quality.

The International Union for Conservation of Nature (IUCN) red list data has included 4 species (*Triaenodes tridontus, Triaenodes phalacris, Rhyacophila amabilis, Hydropsyche tobiasi*) as extinct and 01 species (*Limnephilus atlanticus*) in Near Threatened category. All these species had not been reported from India. The drivers behind the extinction and decline of

Rhyacophilidae of India
Sajad Hussain Parey, Tabraq Ali, and Manpreet Singh Pandher

ISBN 978-981-4968-74-4 (Hardcover), 978-1-003-41095-9 (eBook)
www.jennystanford.com

these species are reported to be agricultural effluents, climate change, invasive species, volcanoes, habitat alteration, droughts, drainage of wetlands and urbanization.

Some of the major threats to the distribution of Trichopteran insects are discussed in the following sections.

9.1 Habitat Loss

Freshwater ecosystems have been lost significantly during the previous decades due to habitat loss, fragmentation, and degradation. The number of surviving free-flowing river stretches around the world is expected to drop by 21% as a result of continued investment in hydropower electricity generation (IUCN data, 2022). The Amazon in South America, the Ganges in India, and the Yangtze in China are the focus of most of these processes, although many of the freshwaters of the globe are also under threat. River drainage and flood control plans that involve straightening and enlarging watercourses also result in a reduction in habitat diversity, both on the banks and in the aquatic environment, as well as changes in environmental variables like water depth and temperature.

9.2 Water Pollution

Fertilizers and pesticides are being added to the stream, and rain may wash an overabundance of sediments into the stream, making agricultural areas particularly problematic. Because these sediments blanket the streambed and darken the water, Trichopteran and other aquatic insects find it difficult to breathe, hunt, and find shelter. Changing the nature of the habitat around the stream can also have an impact on the stream's ecosystem directly and the distribution of Caddisflies indirectly. When a forest is cleared and leaf litter is no longer introduced into the system, there may be less decomposing organic matter in the stream, causing the natural flow of the food chain to be disrupted.

9.3 Industrial Pollution

Man-made activities such as industrial effluents, agricultural operations, urban waste management difficulties, and increased urbanization have all posed severe risks to the freshwater ecosystem in recent years (Zhu et al., 2018). Furthermore, changes in abiotic elements such as precipitation and temperature levels as a result of climate change have impacted the regular function of aquatic ecosystems, including reproduction and feeding. These levels of pollution have also had an impact on the habitats of aquatic flora and fauna (Schmeller et al., 2018). Girgin et al. (2010) investigated the relationship between some aquatic insect orders (Ephemeroptera, Plecoptera, Trichoptera, and Odonata) and heavy metals (cadmium, lead, copper, zinc, nickel, iron, and manganese) and boron, and found that if heavy metal contamination was high, these insect orders may be absent in aquatic conditions (Fig. 9.1).

9.4 Climate Change and Global Warming

One of the biggest long-term risks to biodiversity is climate change, which is well-acknowledged. Temperatures are expected to climb, and extreme weather occurrences will become more frequent, according to recent estimates. Higher water temperatures and changed hydrology will inevitably have an influence on freshwater invertebrate communities. Warm-loving species will expand their range, whereas cold-loving species will retreat northwards and uphill. Climate change is projected to have the greatest influence on Southern Europe, according to a review of European Trichoptera species traits, with up to 30% of the fauna in the Iberic-Macaronesian region potentially endangered by climate change (IUCN data, 2022). Many aquatic insects that are restricted to cold water streams are now absent at many of their original sites at lower altitudes and there are certain shreds of evidence that these insects are being pushed further and further upstream as water temperatures rise.

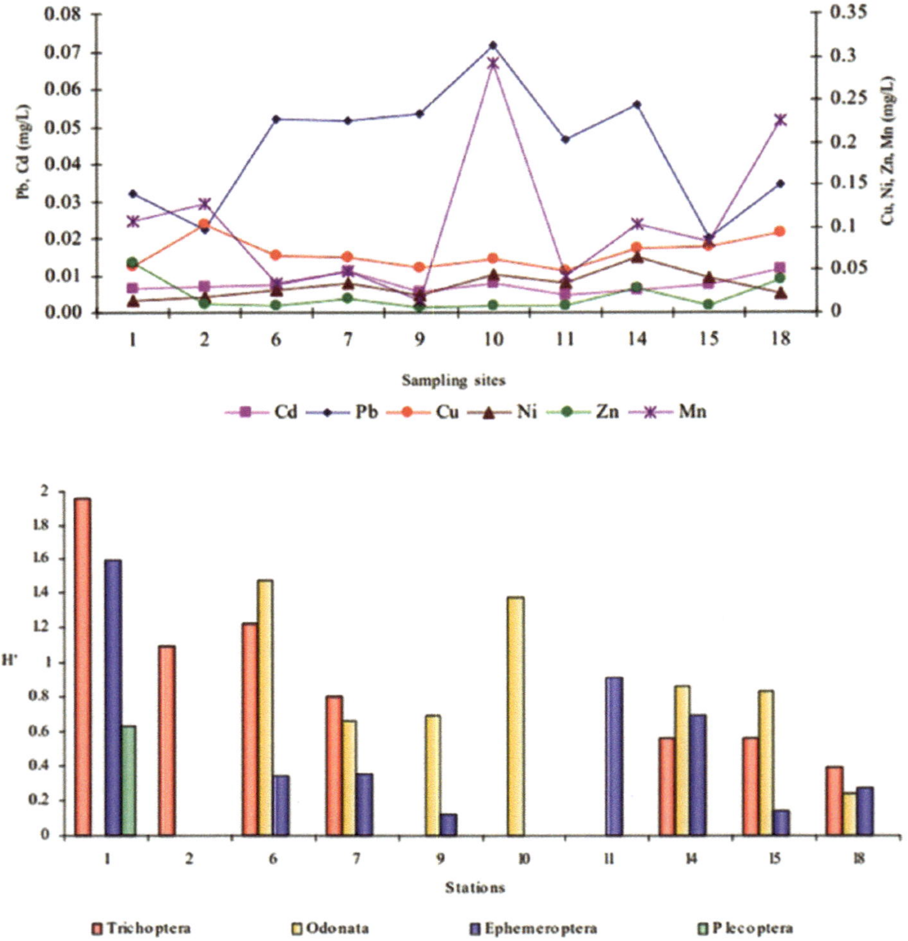

Figure 9.1 Relationship between aquatic insects and heavy metals (Girgin et al., 2010).

9.5 Conservation Approaches

Although no major efforts have been made for the conservation of aquatic insects, scientists very recently raised the issue of declining population of insects and their conservation. The decline in the population of insects implies not only less abundance but also the restricted geographical distribution of species. Almost all the blame for the biodiversity loss falls on the man-made activities such as deforestation, industrialization, agriculture intensification and urbanization, which collectively claimed a 30–50% occupation of natural ecosystems in the 20th century (Vitousek et al., 1997) (Fig. 9.2A–B).

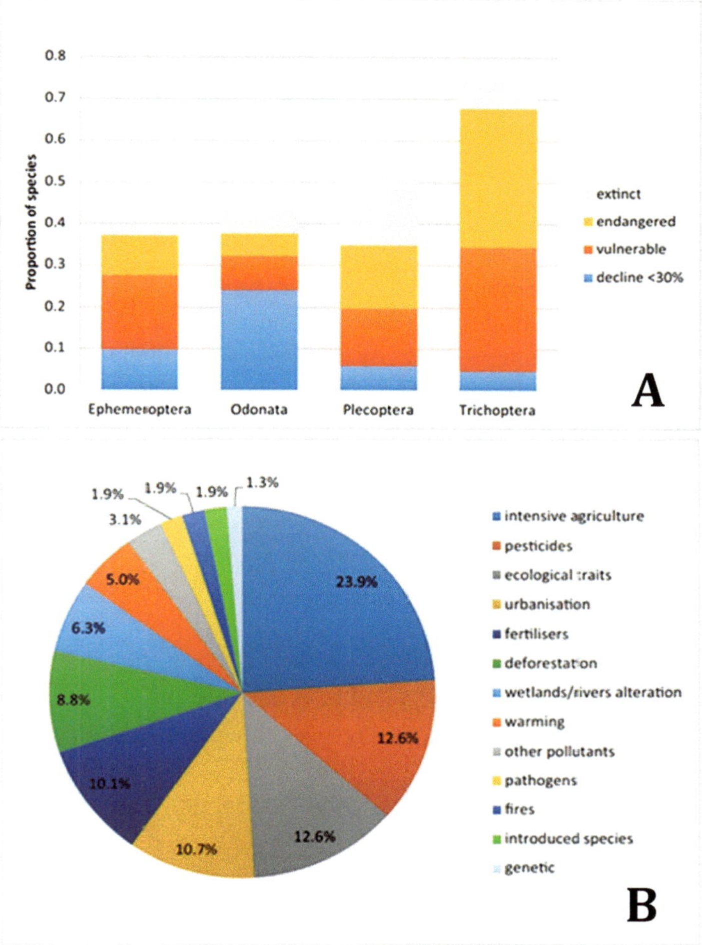

Figure 9.2 (A) Proportion of insect species declining or locally extinct, as per IUCN criteria: vulnerable species (>30% decline), endangered species (>50% decline) and extinct (not recorded for >50 years) in aquatic taxa. (B) Major factors related to insects decline (Francisco Sánchez-Bayo & Kris A. G. Wyckhuys, 2019).

As the major source of destruction is man himself, we may check the risks of the extinction of insects. The following approaches may find a clinical role in the near future as well as in the present scenario.

9.5.1 Ecological Approach

Species traits, life-cycle patterns, and ecological preferences of the given species may be studied for a better understanding of their conservation. The conservation of the ecosystem is the need of the hour which in turn benefits the conservation of insects. At the regional level, headwater stream protection, as well as high-elevation reservoirs for endemic insects, are essential. Habitat fragmentation is another issue that must be halted in which native plants are cleared for human activities such as agriculture, rural development, urbanization and the creation of hydroelectric reservoirs. Avoidance of channelization is very much needed as it reduces insects' diversity and causes hydrological drought. Other insect conservation strategies include: taxonomy and curation of insects, designing sampling protocols, collecting and recording insects and measuring environmental variables.

9.5.2 Biotechnological Approaches

- Methods for quick identification.
- Life stage interactions and DNA barcodes.
- Phylogenetic and phylogeographic investigations at the molecular level.
- Evolutionary lineages and endemic taxa-origin and diversification.
- Conservation of "cryptic species" to protect hidden subspecific diversity.

9.5.3 Sociological and Legal Approaches

- Instilling conservation ethics among the general public and students.

- Raising public awareness about anthropogenic effects, climate change, habitat fragmentation and loss, and the resulting "extinction crisis."
- Increased inter-institutional collaboration in research.
- Relaxation in Legal and regulatory barriers to the open and fair interchange of specimens and genetic material for non-profit academic research.
- Development and execution of an integrated plan for aquatic insect conservation as part of zoo benthic diversity conservation.

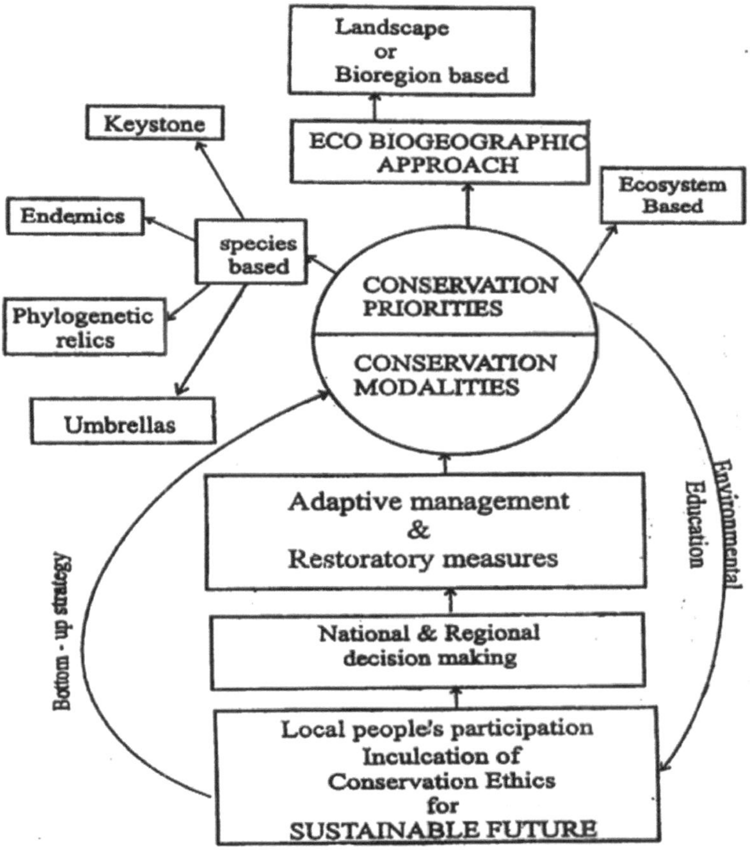

Figure 9.3 Flowchart depicting various approaches towards the conservation of insects.

Acknowledgment

We are highly grateful to people of different regions in India who accompanied us during survey cum collection tours across the Indian Himalayan belt from 2008 to 2022. Without their cooperation, this work would have been very difficult.

We are indebted to our PhD supervisor Prof. M. S. Saini of the Department of Zoology & Environmental Sciences, Punjabi University, Patiala, for introducing us to Indian caddisflies. His mentoring qualities, support and intellectual teaching and learning helped us to groom in the best possible and professional way.

The first author is thankful to Prof. Akbar Masood, Hon'ble Vice Chancellor of BGSB University, Rajouri, for his support and for providing necessary research facilities in the department.

Our work on Indian caddisflies was not possible without the cordial help and support of Dr. John. C. Morse of Clemson University, South Carolina, USA. He provided all the literature on caddisflies through hard copies, emails and hard drives, which ultimately enriched our libraries at both personal and institutional levels.

We are thankful to various funding agencies, including the University Grants Commission (UGC), New Delhi, and the Science and Engineering Research Board (SERB), New Delhi, for providing research funding for working on morphological, molecular and ecological studies of Indian caddisflies.

Last but not the least, we are grateful to our family members who always supported and firmly backed us during these years. They sacrificed days and months when we usually remained in the Indian Himalayan belt for months for the collection of caddisflies.

Sajad H. Parey
Tabraq Ali
Manpreet S. Pandher

Bibliography

Ali, T., Parey, S. H., Pandher, M. S. and Saini, M. S. (2020). Checklist of the caddisfly family Rhyacophilidae (Insecta: Trichoptera) in India. *Insecta Mundi* **0809**, 1–17.

Arnab, B., Indrani, S., Siddartha, D. and Sheela, R. (2018). Community structure of benthic macroinvertebrate fauna of river Ichamati, India. *Journal of Threatened Taxa*, **10**, 12044–12055.

Banks, N. (1931). Some oriental neuropteroid insects. *Psyche* **38**, 56–70.

Banks, N. (1940). Report on certain groups of neuropteroid insects from Szechwan, China. *Proceedings of the United States National Museum* **88**, 173–220.

Barbour, M. T., Gerritsen, J., Snyder, B. D. and Stribling, J. B. (1999). *Rapid Bioassessment Protocols for Use in Streams and Wadeable Rivers: Periphyton, Benthic Macroinvertebrates and Fish*, 2nd ed.; EPA 841-B-99-002; U.S.

Blahnik, R. J. and Holzenthal, R. W. (2004). Collection and curation of Trichoptera, with emphasis on pinned material. *Nectopsyche, the Neotropical Trichoptera Newsletter* **1**, 8–20.

Blahnik, R. J., Holzenthal, R. W. and Prather, A. (2007). The lactic acid method for clearing Trichoptera genitalia. pp. 9–14. In Bueno-Soria, J., Barba-Alvarez, R., Armitage, B. (eds.). *Proceedings of the XIIth International Symposium on Trichoptera*. The Caddis Press; Columbus, OH. xii + 378 p.

Bouchard, Jr. R. (2004). Guide to aquatic macroinvertebrates of the upper midwest waters. *Limnology in Brazil* 365–371.

Calor, A. and Mariano, R. (2009). UV light pan traps for collecting aquatic insects. *EntomoBrasilis* **2**, 1–10. 10.12741/ebrasilis. v5i2.187.

Benchamin, D., Sreejai, R. and Kurup, B. S. (2020). Effect of water quality on caddisflies (Trichoptera) in Kallada river, Kerala, India. *Current Science*, **119**(11), 1845–1849.

de Moor, F. C. and Ivanov, V. D. (2008). Global diversity of Caddisflies (Trichoptera: Insecta) in freshwater. *Hydrobiologia* **595**, 393–407.

DeWalt, R. E., Resh, V. H. and Hilsenhoff, W. L. (2010). Diversity and classification of insects and collembola. In *Ecology and Classification of North American Freshwater Invertebrates* (pp. 587–657). Elsevier Inc. https://doi.org/10.1016/B978-0-12-374855-3.00016-9.

Dohet, A. (2002). Are caddisflies an idea group for the biological assessment of water quality in streams? In W. Mey (ed.), *Proceedings of the 10th International Symposium on Trichoptera* (pp. 507–520). Keltern, Germany: Goecke & Evers.

Flint, O. S. Jr., Holzenthal, R. W. and Harris, S. C. (1999). Catalog of the Neotropical Caddisflies (Insecta: Trichoptera). Ohio Biological Survey, Columbus, Ohio, 239 p.

Sánchez-Bayo, F. and Wyckhuys, K. A. G. (2019). Worldwide decline of the entomofauna: A review of its drivers. *Biological Conservation* **232**, 8–27.

Girgin, Z., Kazanci, N. and Dügel, M. (2010). Relationship between aquatic insects and heavy metals in an urban stream using multivariate techniques. *International Journal of Environmental Science & Technology* **7**(4), 653–664.

Hickin, N. E. (1967). *Caddis Larvae: Larvae of the British Trichoptera,* London: Hutchinson.

Holzenthal, R. W., Blahnik, R. J., Prather, A. and Kjer, K. M. (2007). Order Trichoptera Kirby 1813 (Insecta), caddisflies. *Zootaxa* **1668**, 639–698.

Holzenthal, R. W. (2009). *Encyclopedia of Inland Waters: Trichoptera (Caddisflies)*, pp. 456–467. doi:10.1016/B978-012370626-3.00189-7.

Kaur, L. and Saini, M. S. (2012). Two new species of *Rhyacophila* Pictet (Trichoptera, Rhyacophilidae) from India. *Acta Zoologica Academiae Scientiarum Hungaricae* **58**, 211–215.

Kaur, L. and Saini, M. S. (2013). *Rhyacophila kashmirensis* sp. n. (Trichoptera: Rhyacophilidae) from India. *Polish Journal of Entomology* **82**, 109–112.

Kaur, L. and Saini, M. S. (2015). A new species of the genus *Himalopsyche* (Trichoptera, Rhyacophilidae) with keys to and catalogue of Indian Species. *Vestnikzoologii* **49**, 3–12.

Kimmins, D. E. (1952). New species and a new genus of the subfamily Rhyacophilinae. Indian caddis flies VI. *Annals and Magazine of Natural History (Series 12)* **5**, 347–361.

Kimmins, D. E. (1953). Entomological results from the Swedish Expedition 1934 to Burma and British India V. Trichoptera. The genus *Rhyacophila* Pictet (Fam. Rhyacophilidae). *Arkiv för Zoologi* **2**, 505–555.

Kirby, W. (1813). Strepsiptera, a new order of insects proposed; and the characters of the order, with those of its genera, laid down. *Transactions of the Linnean Society of London* **11**, 86–122.

Kiss, O. (2013). Three new species of *Rhyacophila* (Trichoptera, Rhyacophilidae) from Asia. *Acta Zoologica Academiae Scientiarum Hungaricae* **59**, 13–29.

Klaassen, C. D. (ed.) (2019). *Casarett & Doull's Toxicology: The Basic Science of Poisons*, 9th ed.; McGraw-Hill Education: New York, NY, USA. ISBN 978-1259863745.

Leprieur, F., Beauchard, O., Blanchet, S., Oberdorff, T. and Brosse, S. (2008). Fish invasions in the world's river systems when natural processes are blurred by human activities. *PLoS Biology* **6**, 28.

Malicky, H. (1987). The first European *Apataniana* (Trichoptera, Limnephilidae): *Apatanianahellenica* sp. n, from Greece. *Aquatic Insects* **9**, 210.

Malicky, H. (1997). Weitere neue Köcherfliegen-Arten (Trichoptera) aus Asien. *Linzer Biologische Beiträge* **29**, 217–238.

Martynov, A. V. (1927). Contributions to the aquatic entomofauna of Turkestan I. Trichoptera, Annulipalpia. *Annuaire du Musée Zoologique de l'Académie Impérialedes Sciences de Saint Pétersbourg* **28**, 162–193.

Martynov, A. V. (1930). On the Trichopterous fauna of China and Eastern Tibet. *Proceedings of the Zoological Society of London* **5**, 65–112.

Martynov, A. V. (1935). On a collection of Trichoptera from the Indian Museum, Part I, Annulipalpia. *Records of the Indian Museum* **37**, 93–209.

Martynov, A. V. (1936). On a collection of Trichoptera from the Indian Museum, Part II, Integripalpia. *Records of the Indian Museum, Calcutta* **38**, 239–306.

Morse, J. C. (2017). Biodiversity of aquatic insects. In *Insect Biodiversity: Science and Society*, 2nd ed.; Foottit, R. G., Adler, P. H., eds.; John Wiley & Sons: Hoboken, NJ, USA. Vol. 1, pp. 205–227, ISBN 9781118945537.

Morse, J. C., Frandsen, P. B., Wolfram, G. and Thomas, J. A. (2019). Diversity and ecosystem services of Trichoptera. *Insects* **10**, 125. 10.3390/insects10050125.

Morse, J. C. (ed.) (2022). Trichoptera World Checklist. http://entweb.clemson.edu/database/trichopt/index.htm [Accessed 02 March 2022].

Morton, K. J. (1900). Descriptions of new species of Oriental Rhyacophilidae. *Transactions of the Royal Entomological Society of London* **1900**, 1–7.

Nielsen, A. (1957). A comparative study of genital segments and their appendages in male Trichoptera. *Det Kongelige Danske Videnskabemes Selskab Biologiske Sknfter* **8**(5), 159.

Nielsen, A. (1980). A comparative study of genital segments and the genital chambers in female Trichoptera. *Det Kongelige Danske Videnskabfrences Selskeb Biologiske Skrifter* **23**(1), 1–200.

Pictet, F. J. (1834). *Recherches pour servir à l'histoire et l'anatomie des Phryganides.* A. Cherbuliez, Geneva, plates 1–20, 235 p.

Ross, H. H. (1944). The Caddisflies or Trichoptera of Illinios. *Bulletin of the Illinois National History Survey* **23**, 1–326.

Ross, H. H. (1956). Evolution of the mountain caddisflies. University of Illinois Press; Urbana, Illinois, 213 p.

Saini, M. S., Mandeep, K. and Bajwa, P. K. (2001). An updated checklist of the Indian Trichoptera along with an illustrated key to its families. *Records of the Zoological Survey of India* **99**, 201–256.

Saini, M. S. and Lakhwinder, K. (2011). Revised phylogenetic analysis of Indian species of genus *Himalopsyche* Banks (Trichoptera: Spicipalpia: Rhyacophilidae). *Halteres* **3**, 26–29.

Saini, M. S., Cheema, L. K. and Bajwa, P. K. (2012). Two new species of the genus *Rhyacophila* Pictet (Trichoptera: Rhyacophilidae) from India. *Aquatic Insects* **34**, 223–227.

Schmeller, D. S., Loyau, A., Bao, K., Brack, W., Chatzinotas, A., De Vleeschouwer, F. and Vredenburg, V. T. (2018). People, pollution and pathogens–Global change impacts in mountain freshwater ecosystems. *Science of the Total Environment* **622**, 756–763.

Schmid, F. (1959). Trichoptères du Pakistan, II. *Tijdschrift voor Entomologie* **102**, 231–253.

Schmid, F. (1963). Quelques *Himalopsyche* indiènnes (Trichoptères, Rhyacophilidae). *Bonner Zoologische Beiträge* **14**, 206–223.

Schmid, F. (1970). Le genre *Rhyacophila* et la famille des Rhyacophilidae (Trichoptera). *Memoires de la Société Entomologique du Canada* **66**, 1–230.

Schmid, F. (1989). Les hydrobiosides (Trichoptera, Annulipalpia). *Bulletin de l'Institute Royal des Sciences Naturelles de Belgique, Entomologie* **59**, 1–154.

Schmid, F. and Botosaneanau, L. (1966). Le genre *Himalopsyche* Banks (Trichoptera, Rhyacophilidae). *Extrait des Annalés de la Société entomologique du Québec* **11**, 123–176.

Schmid, F. (1984). Essai d'evaluation de la faune mondiale des trichoptères (abstract), p. 337 in Morse, J. C. (ed.) *Proceedings of the 4th International Symposium on Trichoptera*. The Hague, Dr. W. Junk.

Vitousek, P. M., Mooney, H. A., Lubchenco, J. and Melillo, J. M. (1997). Human domination of Earth's ecosystems. Science **277**, 494–499.

Weaver, J. S., III. and Morse, J. C. (1986). Evolution of feeding and case-making behavior in Trichoptera. *Journal of the North American Benthological Society* **5**, 150–158.

Wheeler, W. C., Whiting, M., Wheeler, Q. D. and Carpenter, J. M. (2001). The phylogeny of the extant hexapod orders. *Cladistics* **17**, 113–169.

Wiggins, G. B. (1982). Trichoptera, pp. 599–612 in Parker, S. P. (ed.) *Synopsis and Classification of Living Organisms*. McGraw Hill Book Company, New York.

Wiggins, G. B. (1996). *Larvae of the North American Caddisfly Genera (Trichoptera)* (2nd ed.). Toronto, Canada: The University of Toronto Press.

Wissinger, S. A., Eldermire, C. and Whissel, J. C. (2004). The role of larval cases in reducing aggression and cannibalism among caddisflies in temporary wetlands. *Wetlands* **24**(4), 777–783.

Zhu, S., Zhang, Z. and Zagar, D. (2018). Mercury transport and fate models in aquatic systems: A review and synthesis. *Science of the Total Environment* **639**, 538–549.

Abbreviations

AB:	Apical Band
aed.:	Aedeagus (phallus)
an.:	Basal ring
AP:	Apodeme
app. d.:	Dorsal appendage of phallic apparatus
app. pr.:	Preanal appendage
AS:	Anal Sclerite
A1-A2:	Anal Veins
BAP:	Basal Apodeme
b. c.:	Copulatory bursa
BMNH:	British Natural History Museum
br. d.:	Dorsal branch of aedeagus (phallus)
br. d. X:	Dorsal branch of Xth segment
br.l.:	Lateral arms of apical band
BS:	Basal segment of inferior appendage
BS:	Basal sleeve of phallus
b. t.:	Tergal band
C:	Costa
c.:	Cerques
corps X^e:	Body of Xth segment
Cu_1-Cu_2:	Cubital Veins
DA:	Dorsal Appendage
DAP:	Dorsal Apodeme
D. IX:	Dorsal Apical lobe of segment IX
END:	Endotheca
end.:	Endotheca
HPX:	Horizontal point of segment X
im:	Inter-segmental membrane
INF:	Inferior Appendage
INP:	Intermediate Process
l. IX:	Apical dorsal lobe of IXth segment
l.v:	Ventral love of phallus
LA:	Lateral arm of ventral lobe of phallicata

LAAB:	Lateral arm of apical band
lobe ap.:	Apical lobe of Xth segment
LP:	Labial Palp
LPR:	Lateral process of dorsal apical lobe of segment IX
MAB:	Median arm of apical band
MDZPUP:	Museum of Department of Zoology & Environmental Sciences
MP:	Maxillary Palp
MPR:	Median process of dorsal apical lobe of segment IX
MST:	Mesoscutellum
M1-M4:	Medial Veins
OCL:	Ocelli
P:	Pedicel
P:	Phallus
PAR:	Paramere
par.:	Paramere
PH:	Phallicata
PHT:	Phallotheca
phall.:	Phallotheca
pl.v.:	Ventral plate of phallus
p. lat.:	Lateral pieces
p. t.:	Terminal pieces
PNSW:	Pronotal Setal Wart
PP:	Posterior Process
PR:	Preanal Appendage
PS:	Phallus sheath
PSP:	Processus Spermathecae
rac.:	Root of anal sclerite
ROM:	Royal Ontario Museum
Rs:	Radial sector
R1-R5:	Radial Veins
R:	Radial cross-vein
r-m:	Radio-median cross-vein
S:	Scape
S:	Scoop or lower part of Xth segment
s. a.:	Anal sclerite
SC:	Subcosta
T:	Tendon

TB:	Tergal Band
TEG:	Tegula
ten.:	Tenons of phallotheca
tend.:	Tendons from inferior appendage
TPHT:	Tenons of phallotheca
TS:	Terminal segment of inferior appendage
U:	Apical band
UV:	Ultraviolet
IXe:	IXth segment
VIIIe t.:	VIIIth tergite
VIIIe st.:	VIIIth sternite
IXe t.:	IXth tergite
IXe st.:	IXth sternite
X^{e} t.:	Xth tergite
XI$^{e:}$	XIth segment
VL:	Ventral Lobe
VLPH:	Ventral lobe of Phallicata
VPX:	Vertical plate of segment X
ZSI:	Zoological Survey of India
V:	Segment V
VI:	Segment VI
VII:	Segment VII
VIII:	Segment VIII
IX:	Segment IX
X:	Segment X
XI:	Segment XI
10T:	10th tergite
8–10:	8th to 10th segments
10S:	10th sternite

Index